Glencoe McGraw-Hill

Homework Practice Workbook

Algebra 1

Mc
Graw
Hill **Glencoe**

To the Student

This *Homework Practice Workbook* gives you additional problems for the concept exercises in each lesson. The exercises are designed to aid your study of mathematics by reinforcing important mathematical skills needed to succeed in the everyday world. The materials are organized by chapter and lesson, with one Practice worksheet for every lesson in *Glencoe Algebra 1*.

To the Teacher

These worksheets are the same ones found in the Chapter Resource Masters for *Glencoe Algebra 1*. The answers to these worksheets are available at the end of each Chapter Resource Masters booklet.

The McGraw-Hill Companies

 Glencoe

Send all inquiries to:
Glencoe/McGraw-Hill
8787 Orion Place
Columbus, OH 43240

ISBN 13: 978-0-07-890836-1
ISBN 10: 0-07-890836-1

Homework Practice Workbook, Algebra 1

Printed in the United States of America

20 MAL 14 13

Contents

Lesson/Title	Page	Lesson/Title	Page

iv

1-1 Skills Practice

Variables and Expressions

Write a verbal expression for each algebraic expression.

1. $9a^2$

2. 5^2

3. $c + 2d$

4. $4 - 5h$

5. $2b^2$

6. $7x^3 - 1$

7. $p^4 + 6r$

8. $3n^2 - x$

Write an algebraic expression for each verbal expression.

9. the sum of a number and 10

10. 15 less than k

11. the product of 18 and q

12. 6 more than twice m

13. 8 increased by three times a number

14. the difference of 17 and 5 times a number

15. the product of 2 and the second power of y

16. 9 less than g to the fourth power

1-1 Practice

Variables and Expressions

Write a verbal expression for each algebraic expression.

1. $23f$

2. 7^3

3. $5m^2 + 2$

4. $4d^3 - 10$

5. $x^3 \cdot y^4$

6. $b^2 - 3c^3$

7. $\dfrac{k^5}{6}$

8. $\dfrac{4n^2}{7}$

Write an algebraic expression for each verbal expression.

9. the difference of 10 and u

10. the sum of 18 and a number

11. the product of 33 and j

12. 74 increased by 3 times y

13. 15 decreased by twice a number

14. 91 more than the square of a number

15. three fourths the square of b

16. two fifths the cube of a number

17. BOOKS A used bookstore sells paperback fiction books in excellent condition for $2.50 and in fair condition for $0.50. Write an expression for the cost of buying x excellent-condition paperbacks and f fair-condition paperbacks.

18. GEOMETRY The surface area of the side of a right cylinder can be found by multiplying twice the number π by the radius times the height. If a circular cylinder has radius r and height h, write an expression that represents the surface area of its side.

 Glencoe Algebra 1

1-2 Skills Practice

Order of Operations

Evaluate each expression.

1. 8^2

2. 3^4

3. 5^3

4. 3^3

5. $(5 + 4) \cdot 7$

6. $(9 - 2) \cdot 3$

7. $4 + 6 \cdot 3$

8. $12 + 2 \cdot 2$

9. $(3 + 5) \cdot 5 + 1$

10. $9 + 4(3 + 1)$

11. $30 - 5 \cdot 4 + 2$

12. $10 + 2 \cdot 6 + 4$

13. $14 \div 7 \cdot 5 - 3^2$

14. $4[30 - (10 - 2) \cdot 3]$

15. $5 + [30 - (6 - 1)^2]$

16. $2[12 + (5 - 2)^2]$

Evaluate each expression if $x = 6$, $y = 8$, and $z = 3$.

17. $xy + z$

18. $yz - x$

19. $2x + 3y - z$

20. $2(x + z) - y$

21. $5z + (y - x)$

22. $5x - (y + 2z)$

23. $x^2 + y^2 - 10z$

24. $z^3 + (y^2 - 4x)$

25. $\dfrac{y + xz}{2}$

26. $\dfrac{3y + x^2}{z}$

1-2 Practice

Order of Operations

Evaluate each expression.

1. 11^2
2. 8^3
3. 5^4

4. $(15 - 5) \cdot 2$
5. $9 \cdot (3 + 4)$
6. $5 + 7 \cdot 4$

7. $4(3 + 5) - 5 \cdot 4$
8. $22 \div 11 \cdot 9 - 3^2$
9. $6^2 + 3 \cdot 7 - 9$

10. $3[10 - (27 \div 9)]$
11. $2[5^2 + (36 \div 6)]$
12. $162 \div [6(7 - 4)^2]$

13. $\dfrac{5^2 \cdot 4 - 5 \cdot 4^2}{5(4)}$
14. $\dfrac{(2 \cdot 5)^2 + 4}{3^2 - 5}$
15. $\dfrac{7 + 3^2}{4^2 \cdot 2}$

Evaluate each expression if $a = 12$, $b = 9$, and $c = 4$.

16. $a^2 + b - c^2$
17. $b^2 + 2a - c^2$

18. $2c(a + b)$
19. $4a + 2b - c^2$

20. $(a^2 \div 4b) + c$
21. $c^2 \cdot (2b - a)$

22. $\dfrac{bc^2 + a}{c}$
23. $\dfrac{2c^3 - ab}{4}$

24. $2(a - b)^2 - 5c$
25. $\dfrac{b^2 - 2c^2}{a + c - b}$

26. **CAR RENTAL** Ann Carlyle is planning a business trip for which she needs to rent a car. The car rental company charges $36 per day plus $0.50 per mile over 100 miles. Suppose Ms. Carlyle rents the car for 5 days and drives 180 miles.

   a. Write an expression for how much it will cost Ms. Carlyle to rent the car.

   b. Evaluate the expression to determine how much Ms. Carlyle must pay the car rental company.

27. **GEOMETRY** The length of a rectangle is $3n + 2$ and its width is $n - 1$. The perimeter of the rectangle is twice the sum of its length and its width.

   a. Write an expression that represents the perimeter of the rectangle.

   b. Find the perimeter of the rectangle when $n = 4$ inches.

1-3 Skills Practice

Properties of Numbers

Evaluate each expression. Name the property used in each step.

1. $7(16 \div 4^2)$

2. $2[5 - (15 \div 3)]$

3. $4 - 3[7 - (2 \cdot 3)]$

4. $4[8 - (4 \cdot 2)] + 1$

5. $6 + 9[10 - 2(2 + 3)]$

6. $2(6 \div 3 - 1) \cdot \frac{1}{2}$

7. $16 + 8 + 14 + 12$

8. $36 + 23 + 14 + 7$

9. $5 \cdot 3 \cdot 4 \cdot 3$

10. $2 \cdot 4 \cdot 5 \cdot 3$

1-3 Practice

Properties of Numbers

Evaluate each expression. Name the property used in each step.

1. $2 + 6(9 - 3^2) - 2$

2. $5(14 - 39 \div 3) + 4 \cdot \frac{1}{4}$

Evaluate each expression using properties of numbers. Name the property used in each step.

3. $13 + 23 + 12 + 7$

4. $6 \cdot 0.7 \cdot 5$

5. SALES Althea paid $5.00 each for two bracelets and later sold each for $15.00. She paid $8.00 each for three bracelets and sold each of them for $9.00.

   a. Write an expression that represents the profit Althea made.

   b. Evaluate the expression. Name the property used in each step.

6. SCHOOL SUPPLIES Kristen purchased two binders that cost $1.25 each, two binders that cost $4.75 each, two packages of paper that cost $1.50 per package, four blue pens that cost $1.15 each, and four pencils that cost $.35 each.

   a. Write an expression to represent the total cost of supplies before tax.

   b. What was the total cost of supplies before tax?

1-4 Skills Practice
The Distributive Property

Use the Distributive Property to rewrite each expression. Then evaluate.

1. $4(3 + 5)$

2. $2(6 + 10)$

3. $5(7 - 4)$

4. $(6 - 2)8$

5. $5 \cdot 89$

6. $9 \cdot 99$

7. $15 \cdot 104$

8. $15\left(2\frac{1}{3}\right)$

Use the Distributive Property to rewrite each expression. Then evaluate.

9. $(a + 7)2$

10. $7(h - 10)$

11. $3(m + n)$

12. $2(x - y + 1)$

Simplify each expression. If not possible, write *simplified*.

13. $2x + 8x$

14. $17g + g$

15. $2x^2 + 6x^2$

16. $7a^2 - 2a^2$

17. $3y^2 - 2y$

18. $2(n + 2n)$

19. $4(2b - b)$

20. $3q^2 + q - q^2$

Write an algebraic expression for each verbal expression. Then simplify, indicating the properties used.

21. The product of 9 and t squared, increased by the sum of the square of t and 2

22. 3 times the sum of r and d squared minus 2 times the sum of r and d squared

1-4 Practice

The Distributive Property

Use the Distributive Property to rewrite each expression. Then evaluate.

1. $9(7 + 8)$

2. $7(6 - 4)$

3. $(4 + 6)11$

4. $9 \cdot 499$

5. $7 \cdot 110$

6. $16\left(4\frac{1}{4}\right)$

Use the Distributive property to rewrite each expression. Then simplify.

7. $(9 - p)3$

8. $(5y - 3)7$

9. $15\left(f + \frac{1}{3}\right)$

10. $16(3b - 0.25)$

11. $m(n + 4)$

12. $(c - 4)d$

Simplify each expression. If not possible, write *simplified*.

13. $w + 14w - 6w$

14. $3(5 + 6h)$

15. $12b^2 + 9b^2$

16. $25t^3 - 17t^3$

17. $3a^2 + 6a + 2b^2$

18. $4(6p + 2q - 2p)$

Write an algebraic expression for each verbal expression. Then simplify, indicating the properties used.

19. 4 times the difference of f squared and g, increased by the sum of f squared and $2g$

20. 3 times the sum of x and y squared plus 5 times the difference of $2x$ and y

21. DINING OUT The Ross family recently dined at an Italian restaurant. Each of the four family members ordered a pasta dish that cost $11.50, a drink that cost $1.50, and dessert that cost $2.75.

   a. Write an expression that could be used to calculate the cost of the Ross' dinner before adding tax and a tip.

   b. What was the cost of dining out for the Ross family?

1-5 Skills Practice

Equations

Find the solution of each equation if the replacement sets are $A = \{4, 5, 6, 7, 8\}$ and $B = \{9, 10, 11, 12, 13\}$.

1. $5a - 9 = 26$ **2.** $4a - 8 = 16$

3. $7a + 21 = 56$ **4.** $3b + 15 = 48$

5. $4b - 12 = 28$ **6.** $\dfrac{36}{b} - 3 = 0$

Find the solution of each equation using the given replacement set.

7. $\dfrac{1}{2} + x = \dfrac{5}{4}; \left\{\dfrac{1}{2}, \dfrac{3}{4}, 1, \dfrac{5}{4}\right\}$ **8.** $x + \dfrac{2}{3} = \dfrac{13}{9}; \left\{\dfrac{5}{9}, \dfrac{2}{3}, \dfrac{7}{9}\right\}$

9. $\dfrac{1}{4}(x + 2) = \dfrac{5}{6}; \left\{\dfrac{2}{3}, \dfrac{3}{4}, \dfrac{5}{4}, \dfrac{4}{3}\right\}$ **10.** $0.8(x + 5) = 5.2; \{1.2, 1.3, 1.4, 1.5\}$

Solve each equation.

11. $10.4 - 6.8 = x$ **12.** $y = 20.1 - 11.9$

13. $\dfrac{46 - 15}{3 + 28} = a$ **14.** $c = \dfrac{6 + 18}{31 - 25}$

15. $\dfrac{2(4) + 4}{3(3 - 1)} = b$ **16.** $\dfrac{6(7 - 2)}{3(8) + 6} = n$

17. SHOPPING ONLINE Jennifer is purchasing CDs and a new CD player from an online store. She pays $10 for each CD, as well as $50 for the CD player. Write and solve an equation to find the total amount Jennifer spent if she buys 4 CDs and a CD player from the store.

18. TRAVEL An airplane can travel at a speed of 550 miles per hour. Write and solve an equation to find the time it will take to fly from London to Montreal, a distance of approximately 3300 miles.

1-5 Practice

Equations

Find the solution of each equation if the replacement sets are $a = \left\{0, \frac{1}{2}, 1, \frac{3}{2}, 2\right\}$ and $b = \{3, 3.5, 4, 4.5, 5\}$.

1. $a + \frac{1}{2} = 1$ **2.** $4b - 8 = 6$ **3.** $6a + 18 = 27$

4. $7b - 8 = 16.5$ **5.** $120 - 28a = 78$ **6.** $\frac{28}{b} + 9 = 16$

Solve each equation.

7. $x = 18.3 - 4.8$ **8.** $w = 20.2 - 8.95$ **9.** $\frac{37 - 9}{18 - 11} = d$

10. $\frac{97 - 25}{41 - 23} = k$ **11.** $y = \frac{4(22 - 4)}{3(6) + 6}$ **12.** $\frac{5(2^2) + 4(3)}{4(2^3 - 4)} = p$

13. TEACHING A teacher has 15 weeks in which to teach six chapters. Write and then solve an equation that represents the number of lessons the teacher must teach per week if there is an average of 8.5 lessons per chapter.

14. CELL PHONES Gabriel pays $40 a month for basic cell phone service. In addition, Gabriel can send text messages for $0.20 each. Write and solve an equation to find the total amount Gabriel spent this month if he sends 40 text messages.

1-6 Skills Practice

Relations

Express each relation as a table, a graph, and a mapping. Then determine the domain and range.

1. {(−1, −1), (1, 1), (2, 1), (3, 2)}

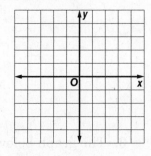

x	y

2. {(0, 4), (−4, −4), (−2, 3), (4, 0)}

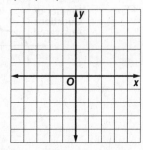

x	y

3. {(3, −2), (1, 0), (−2, 4), (3, 1)}

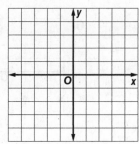

x	y

Identify the independent and dependent variables for each relation.

4. The more hours Maribel works at her job, the larger her paycheck becomes.

5. Increasing the price of an item decreases the amount of people willing to buy it.

1-6 Practice

Relations

1. Express {(4, 3), (−1, 4), (3, −2), (−2, 1)} as a table, a graph, and a mapping. Then determine the domain and range.

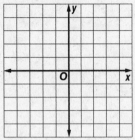

Describe what is happening in each graph.

2. The graph below represents the height of a tsunami (tidal wave) as it approaches shore.

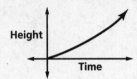

3. The graph below represents a student taking an exam.

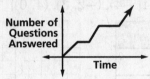

Express the relation shown in each table, mapping, or graph as a set of ordered pairs.

4.

X	Y
0	9
−8	3
2	−6
1	4

5.

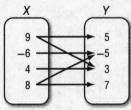

6.

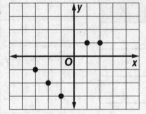

7. **BASEBALL** The graph shows the number of home runs hit by Andruw Jones of the Atlanta Braves. Express the relation as a set of ordered pairs. Then describe the domain and range.

Andruw Jones' Home Runs

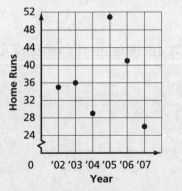

1-7 Skills Practice

Functions

Determine whether each relation is a function. Explain.

1.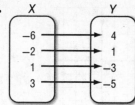

2.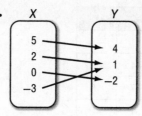

3.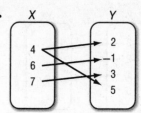

4.
x	y
4	−5
−1	−10
0	−9
1	−7
9	1

5.
x	y
2	7
5	−3
3	5
−4	−2
5	2

6.
x	y
3	7
−1	1
1	0
3	5
7	3

7. {(2, 5), (4, −2), (3, 3), (5, 4), (−2, 5)}

8. {(6, −1), (−4, 2), (5, 2), (4, 6), (6, 5)}

9. $y = 2x - 5$

10. $y = 11$

11.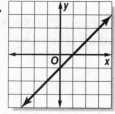

12.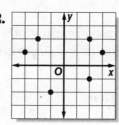

13.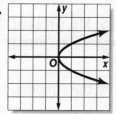

If $f(x) = 3x + 2$ and $g(x) = x^2 - x$, find each value.

14. $f(4)$

15. $f(8)$

16. $f(-2)$

17. $g(2)$

18. $g(-3)$

19. $g(-6)$

20. $f(2) + 1$

21. $f(1) - 1$

22. $g(2) - 2$

23. $g(-1) + 4$

24. $f(x + 1)$

25. $g(3b)$

1-7 Practice

Functions

Determine whether each relation is a function. Explain.

1.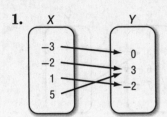

2.

X	Y
1	−5
−4	3
7	6
1	−2

3.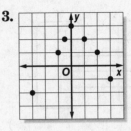

4. {(1, 4), (2, −2), (3, −6), (−6, 3), (−3, 6)}   5. {(6, −4), (2, −4), (−4, 2), (4, 6), (2, 6)}

6. $x = -2$ 7. $y = 2$

If $f(x) = 2x - 6$ and $g(x) = x - 2x^2$, find each value.

8. $f(2)$ 9. $f\left(-\frac{1}{2}\right)$ 10. $g(-1)$

11. $g\left(-\frac{1}{3}\right)$ 12. $f(7) - 9$ 13. $g(-3) + 13$

14. $f(h + 9)$ 15. $g(3y)$ 16. $2[g(b) + 1]$

17. **WAGES** Martin earns $7.50 per hour proofreading ads at a local newspaper. His weekly wage w can be described by the equation $w = 7.5h$, where h is the number of hours worked.

   a. Write the equation in function notation.

   b. Find $f(15)$, $f(20)$, and $f(25)$.

18. **ELECTRICITY** The table shows the relationship between resistance R and current I in a circuit.

Resistance (ohms)	120	80	48	6	4
Current (amperes)	0.1	0.15	0.25	2	3

   a. Is the relationship a function? Explain.

   b. If the relation can be represented by the equation $IR = 12$, rewrite the equation in function notation so that the resistance R is a function of the current I.

   c. What is the resistance in a circuit when the current is 0.5 ampere?

1-8 Skills Practice

Logical Reasoning and Counterexamples

Identify the hypothesis and conclusion of each statement.

1. If it is Sunday, then mail is not delivered.

2. If you are hiking in the mountains, then you are outdoors.

3. If $6n + 4 > 58$, then $n > 9$.

Identify the hypothesis and conclusion of each statement. Then write the statement in if-then form.

4. Martina works at the bakery every Saturday.

5. Ivan only runs early in the morning.

6. A polygon that has five sides is a pentagon.

Determine whether a valid conclusion follows from the statement *If Hector scores an 85 or above on his science exam, then he will earn an A in the class* for the given condition. If a valid conclusion does not follow, write *no valid conclusion* and explain why.

7. Hector scored an 86 on his science exam.

8. Hector did not earn an A in science.

9. Hector scored 84 on the science exam.

10. Hector studied 10 hours for the science exam.

Find a counterexample for each conditional statement.

11. If the car will not start, then it is out of gas.

12. If the basketball team has scored 100 points, then they must be winning the game.

13. If the Commutative Property holds for addition, then it holds for subtraction.

14. If $2n + 3 < 17$, then $n \leq 7$.

1-8 Practice

Logical Reasoning and Counterexamples

Identify the hypothesis and conclusion of each statement.

1. If it is raining, then the meteorologist's prediction was accurate.

2. If $x = 4$, then $2x + 3 = 11$.

Identify the hypothesis and conclusion of each statement. Then write the statement in if-then form.

3. When Joseph has a fever, he stays home from school.

4. Two congruent triangles are similar.

Determine whether a valid conclusion follows from the statement *If two numbers are even, then their product is even* for the given condition. If a valid conclusion does not follow, write *no valid conclusion* and explain why.

5. The product of two numbers is 12.

6. Two numbers are 8 and 6.

Find a counterexample for each conditional statement.

7. If the refrigerator stopped running, then there was a power outage.

8. If $6h - 7 < 5$, then $h \leq 2$.

9. GEOMETRY Consider the statement: If the perimeter of a rectangle is 14 inches, then its area is 10 square inches.

    a. State a condition in which the hypothesis and conclusion are valid.

    b. Provide a counterexample to show the statement is false.

10. ADVERTISING A recent television commercial for a car dealership stated that "no reasonable offer will be refused." Identify the hypothesis and conclusion of the statement. Then write the statement in if-then form.

2-1 Skills Practice

Writing Equations

Translate each sentence into an equation.

1. Two added to three times a number m is the same as 18.

2. Twice a increased by the cube of a equals b.

3. Seven less than the sum of p and t is as much as 6.

4. The sum of x and its square is equal to y times z.

5. Four times the sum of f and g is identical to six times g.

Translate each sentence into a formula.

6. The perimeter P of a square equals four times the length of a side ℓ.

7. The area A of a square is the length of a side ℓ squared.

8. The perimeter P of a triangle is equal to the sum of the lengths of sides a, b, and c.

9. The area A of a circle is pi times the radius r squared.

10. The volume V of a rectangular prism equals the product of the length ℓ, the width w, and the height h.

Translate each equation into a sentence.

11. $g + 10 = 3g$

12. $2p + 4t = 20$

13. $4(a + b) = 9a$

14. $8 - 6x = 4 + 2x$

15. $\frac{1}{2}(f + y) = f - 5$

16. $k^2 - n^2 = 2b$

Write a problem based on the given information.

17. c = cost per pound of plain coffee beans
$c + 3$ = cost per pound of flavored coffee beans
$2c + (c + 3) = 21$

18. p = cost of dinner
$0.15p$ = cost of a 15% tip
$p + 0.15p = 23$

2-1 Practice

Writing Equations

Translate each sentence into an equation.

1. Fifty-three plus four times b is as much as 21.

2. The sum of five times h and twice g is equal to 23.

3. One fourth the sum of r and ten is identical to r minus 4.

4. Three plus the sum of the squares of w and x is 32.

Translate each sentence into a formula.

5. Degrees Kelvin K equals 273 plus degrees Celsius C.

6. The total cost C of gas is the price p per gallon times the number of gallons g.

7. The sum S of the measures of the angles of a polygon is equal to 180 times the difference of the number of sides n and 2.

Translate each equation into a sentence.

8. $r - (4 + p) = \frac{1}{3}r$

9. $\frac{3}{5}t + 2 = t$

10. $9(y^2 + x) = 18$

11. $2(m - n) = x + 7$

Write a problem based on the given information.

12. a = cost of one adult's ticket to zoo
$a - 4$ = cost of one children's ticket to zoo
$2a + 4(a - 4) = 38$

13. c = regular cost of one airline ticket
$0.20c$ = amount of 20% promotional discount
$3(c - 0.20c) = 330$

14. **GEOGRAPHY** About 15% of all federally-owned land in the 48 contiguous states of the United States is in Nevada. If F represents the area of federally-owned land in these states, and N represents the portion in Nevada, write an equation for this situation.

15. **FITNESS** Deanna and Pietra each go for walks around a lake a few times per week. Last week, Deanna walked 7 miles more than Pietra.

   a. If p represents the number of miles Pietra walked, write an equation that represents the total number of miles T the two girls walked.

   b. If Pietra walked 9 miles during the week, how many miles did Deanna walk?

   c. If Pietra walked 11 miles during the week, how many miles did the two girls walk together?

2-2 Skills Practice

Solving One-Step Equations

Solve each equation. Check your solution.

1. $y - 7 = 8$

2. $w + 14 = -8$

3. $p - 4 = 6$

4. $-13 = 5 + x$

5. $98 = b + 34$

6. $y - 32 = -1$

7. $n + (-28) = 0$

8. $y + (-10) = 6$

9. $-1 = t + (-19)$

10. $j - (-17) = 36$

11. $14 = d + (-10)$

12. $u + (-5) = -15$

13. $11 = -16 + y$

14. $c - (-3) = 100$

15. $47 = w - (-8)$

16. $x - (-74) = -22$

17. $4 - (-h) = 68$

18. $-56 = 20 - (-j)$

19. $12z = 108$

20. $-7t = 49$

21. $18f = -216$

22. $-22 = 11v$

23. $-6d = -42$

24. $96 = -24a$

25. $\frac{c}{4} = 16$

26. $\frac{a}{16} = 9$

27. $-84 = \frac{d}{3}$

28. $-\frac{d}{7} = -13$

29. $\frac{t}{4} = -13$

30. $31 = -\frac{1}{6}n$

31. $-6 = \frac{2}{3}z$

32. $\frac{2}{7}q = -4$

33. $\frac{5}{9}p = -10$

34. $\frac{a}{10} = \frac{2}{5}$

2-2 Practice

Solving One-Step Equations

Solve each equation. Check your solution.

1. $d - 8 = 17$

2. $v + 12 = -5$

3. $b - 2 = -11$

4. $-16 = m + 71$

5. $29 = a - 76$

6. $-14 + y = -2$

7. $8 - (-n) = 1$

8. $78 + r = -15$

9. $f + (-3) = -9$

10. $8j = 96$

11. $-13z = -39$

12. $-180 = 15m$

13. $243 = 27r$

14. $\frac{y}{9} = -8$

15. $-\frac{j}{12} = -8$

16. $\frac{a}{15} = \frac{4}{5}$

17. $\frac{g}{27} = \frac{2}{9}$

18. $\frac{q}{24} = \frac{1}{6}$

Write an equation for each sentence. Then solve the equation.

19. Negative nine times a number equals -117.

20. Negative one eighth of a number is $-\frac{3}{4}$.

21. Five sixths of a number is $-\frac{5}{9}$.

22. 2.7 times a number equals 8.37.

23. **HURRICANES** The day after a hurricane, the barometric pressure in a coastal town has risen to 29.7 inches of mercury, which is 2.9 inches of mercury higher than the pressure when the eye of the hurricane passed over.

    a. Write an addition equation to represent the situation.

    b. What was the barometric pressure when the eye passed over?

24. **ROLLER COASTERS** *Kingda Ka* in New Jersey is the tallest and fastest roller coaster in the world. Riders travel at an average speed of 61 feet per second for 3118 feet. They reach a maximum speed of 187 feet per second.

    a. If x represents the total time that the roller coaster is in motion for each ride, write an expression to represent the sitation. (*Hint:* Use the distance formula $d = rt$.)

    b. How long is the roller coaster in motion?

2-3 Skills Practice

Solving Multi-Step Equations

Solve each problem by working backward.

1. A number is divided by 2, and then the quotient is added to 8. The result is 33. Find the number.

2. Two is subtracted from a number, and then the difference is divided by 3. The result is 30. Find the number.

3. A number is multiplied by 2, and then the product is added to 9. The result is 49. What is the number?

4. **ALLOWANCE** After Ricardo received his allowance for the week, he went to the mall with some friends. He spent half of his allowance on a new paperback book. Then he bought himself a snack for $1.25. When he arrived home, he had $5.00 left. How much was his allowance?

Solve each equation. Check your solution.

5. $5x + 3 = 23$

6. $4 = 3a - 14$

7. $2y + 5 = 19$

8. $6 + 5c = -29$

9. $8 - 5w = -37$

10. $18 - 4v = 42$

11. $\frac{n}{3} - 8 = -2$

12. $5 + \frac{x}{4} = 1$

13. $-\frac{h}{3} - 4 = 13$

14. $-\frac{d}{6} + 12 = -7$

15. $\frac{a}{5} - 2 = 9$

16. $\frac{w}{7} + 3 = -1$

17. $\frac{3}{4}q - 7 = 8$

18. $\frac{2}{3}g + 6 = -12$

19. $\frac{5}{2}z - 8 = -3$

20. $\frac{4}{5}m + 2 = 6$

21. $\frac{c - 5}{4} = 3$

22. $\frac{b + 1}{3} = 2$

Write an equation and solve each problem.

23. Twice a number plus four equals 6. What is the number?

24. Sixteen is seven plus three times a number. Find the number.

25. Find two consecutive integers whose sum is 35.

26. Find three consecutive integers whose sum is 36.

2-3 Practice

Solving Multi-Step Equations

Solve each problem by working backward.

1. Three is added to a number, and then the sum is multiplied by 4. The result is 16. Find the number.

2. A number is divided by 4, and the quotient is added to 3. The result is 24. What is the number?

3. Two is subtracted from a number, and then the difference is multiplied by 5. The result is 30. Find the number.

4. **BIRD WATCHING** While Michelle sat observing birds at a bird feeder, one fourth of the birds flew away when they were startled by a noise. Two birds left the feeder to go to another stationed a few feet away. Three more birds flew into the branches of a nearby tree. Four birds remained at the feeder. How many birds were at the feeder initially?

Solve each equation. Check your solution.

5. $-12n - 19 = 77$

6. $17 + 3f = 14$

7. $15t + 4 = 49$

8. $\dfrac{u}{5} + 6 = 2$

9. $\dfrac{d}{-4} + 3 = 15$

10. $\dfrac{b}{3} - 6 = -2$

11. $\dfrac{1}{2}y - \dfrac{1}{8} = \dfrac{7}{8}$

12. $-32 - \dfrac{3}{5}f = -17$

13. $8 - \dfrac{3}{8}k = -4$

14. $\dfrac{r + 13}{12} = 1$

15. $\dfrac{15 - a}{3} = -9$

16. $\dfrac{3k - 7}{5} = 16$

17. $\dfrac{x}{7} - 0.5 = 2.5$

18. $2.5g + 0.45 = 0.95$

19. $0.4m - 0.7 = 0.22$

Write an equation and solve each problem.

20. Seven less than four times a number equals 13. What is the number?

21. Find two consecutive odd integers whose sum is 116.

22. Find two consecutive even integers whose sum is 126.

23. Find three consecutive odd integers whose sum is 117.

24. **COIN COLLECTING** Jung has a total of 92 coins in his coin collection. This is 8 more than three times the number of quarters in the collection. How many quarters does Jung have in his collection?

2-4 Skills Practice

Solving Equations with the Variable on Each Side

Justify each step.

1.
$$4k - 3 = 2k + 5$$

$4k - 3 - 2k = 2k + 5 - 2k$ **a.** _____

$2k - 3 = 5$ **b.** _____

$2k - 3 + 3 = 5 + 3$ **c.** _____

$2k = 8$ **d.** _____

$\dfrac{2k}{2} = \dfrac{8}{2}$ **e.** _____

$k = 4$ **f.** _____

2.
$$2(8u + 2) = 3(2u - 7)$$

$16u + 4 = 6u - 21$ **a.** _____

$16u + 4 - 6u = 6u - 21 - 6u$ **b.** _____

$10u + 4 = -21$ **c.** _____

$10u + 4 - 4 = -21 - 4$ **d.** _____

$10u = -25$ **e.** _____

$\dfrac{10u}{10} = \dfrac{-25}{10}$ **f.** _____

$u = -2.5$ **g.** _____

Solve each equation. Check your solution.

3. $2m + 12 = 3m - 31$

4. $2h - 8 = h + 17$

5. $7a - 3 = 3 - 2a$

6. $4n - 12 = 12 - 4n$

7. $4x - 9 = 7x + 12$

8. $-6y - 3 = 3 - 6y$

9. $5 + 3r = 5r - 19$

10. $-9 + 8k = 7 + 4k$

11. $8q + 12 = 4(3 + 2q)$

12. $3(5j + 2) = 2(3j - 6)$

13. $6(-3v + 1) = 5(-2v - 2)$

14. $-7(2b - 4) = 5(-2b + 6)$

15. $3(8 - 3t) = 5(2 + t)$

16. $2(3u + 7) = -4(3 - 2u)$

17. $8(2f - 2) = 7(3f + 2)$

18. $5(-6 - 3d) = 3(8 + 7d)$

19. $6(w - 1) = 3(3w + 5)$

20. $7(-3y + 2) = 8(3y - 2)$

21. $\dfrac{2}{3}v - 6 = 6 - \dfrac{2}{3}v$

22. $\dfrac{1}{2} - \dfrac{5}{8}x = \dfrac{7}{8}x + \dfrac{7}{2}$

2-4 Practice

Solving Equations with the Variable on Each Side

Solve each equation. Check your solution.

1. $5x - 3 = 13 - 3x$

2. $-4r - 11 = 4r + 21$

3. $1 - m = 6 - 6m$

4. $14 + 5n = -4n + 17$

5. $\frac{1}{2}k - 3 = 2 - \frac{3}{4}k$

6. $\frac{1}{2}(6 - y) = y$

7. $3(-2 - 3x) = -9x - 4$

8. $4(4 - w) = 3(2w + 2)$

9. $9(4b - 1) = 2(9b + 3)$

10. $3(6 + 5y) = 2(-5 + 4y)$

11. $-5x - 10 = 2 - (x + 4)$

12. $6 + 2(3j - 2) = 4(1 + j)$

13. $\frac{5}{2}t - t = 3 + \frac{3}{2}t$

14. $1.4f + 1.1 = 8.3 - f$

15. $\frac{2}{3}x - \frac{1}{6} = \frac{1}{2}x + \frac{5}{6}$

16. $2 - \frac{3}{4}k = \frac{1}{8}k + 9$

17. $\frac{1}{2}(3g - 2) = \frac{g}{2}$

18. $\frac{1}{3}(n + 1) = \frac{1}{6}(3n - 5)$

19. $\frac{1}{2}(5 - 2h) = \frac{h}{2}$

20. $\frac{1}{9}(2m - 16) = \frac{1}{3}(2m + 4)$

21. $3(d - 8) - 5 = 9(d + 2) + 1$

22. $2(a - 8) + 7 = 5(a + 2) - 3a - 19$

23. NUMBERS Two thirds of a number reduced by 11 is equal to 4 more than the number. Find the number.

24. NUMBERS Five times the sum of a number and 3 is the same as 3 multiplied by 1 less than twice the number. What is the number?

25. NUMBER THEORY Tripling the greater of two consecutive even integers gives the same result as subtracting 10 from the lesser even integer. What are the integers?

26. GEOMETRY The formula for the perimeter of a rectangle is $P = 2\ell, + 2w$, where ℓ is the length and w is the width. A rectangle has a perimeter of 24 inches. Find its dimensions if its length is 3 inches greater than its width.

2-5 **Skills Practice**

Solving Equations Involving Absolute Value

Evaluate each expression if $a = 2$, $b = -3$, and $c = -4$.

1. $|a - 5| - 1$

2. $|b + 1| + 8$

3. $5 - |c + 1|$

4. $|a + b| - c$

Solve each equation. Then graph the solution set.

5. $|w + 1| = 5$

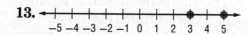

6. $|c - 3| = 1$

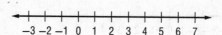

7. $|n + 2| = 1$

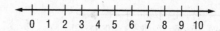

8. $|t + 6| = 4$

9. $|w - 2| = 2$

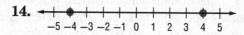

10. $|k - 5| = 4$

Write an equation involving absolute value for each graph.

11.
```
  +--+--+--+--●--+--●--+--+--+--+
 -5 -4 -3 -2 -1  0  1  2  3  4  5
```

12.
```
  +--+--●--+--+--+--●--+--+--+
 -7 -6 -5 -4 -3 -2 -1  0  1  2  3
```

13.
```
  +--+--+--+--+--+--+--●--+--●--+
 -5 -4 -3 -2 -1  0  1  2  3  4  5
```

14.
```
  +--●--+--+--+--+--+--+--+--●--+
 -5 -4 -3 -2 -1  0  1  2  3  4  5
```

2-5 Practice

Solving Equations Involving Absolute Value

Evaluate each expression if $x = -1$, $y = 3$, **and** $z = -4$.

1. $16 - |2z + 1|$

2. $|x - y| + 4$

3. $|-3y + z| - x$

4. $3|z - x| + |2 - y|$

Solve each equation. Then graph the solution set.

5. $|2z - 9| = 1$

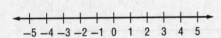

6. $|3 - 2r| = 7$

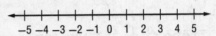

7. $|3t + 6| = 9$

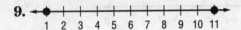

8. $|2g - 5| = 9$

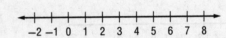

Write an equation involving absolute value for each graph.

9.

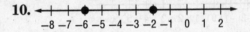

10.

11.

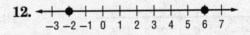

12.

13. FITNESS Taisha uses the elliptical cross-trainer at the gym. Her general goal is to burn 280 Calories per workout, but she varies by as much as 25 Calories from this amount on any given day. Write and solve an equation to find the maximum and minimum number of Calories Taisha burns on the cross-trainer.

14. TEMPERATURE A thermometer is guaranteed to give a temperature no more than 1.2°F from the actual temperature. If the thermometer reads 28°F, write and solve an equation to find the maximum and minimum temperatures it could be.

2-6 Skills Practice

Ratios and Proportions

Determine whether each pair of ratios are equivalent ratios. Write *yes* or *no*.

1. $\dfrac{4}{5}, \dfrac{20}{25}$

2. $\dfrac{5}{9}, \dfrac{7}{11}$

3. $\dfrac{6}{7}, \dfrac{24}{28}$

4. $\dfrac{8}{9}, \dfrac{72}{81}$

5. $\dfrac{7}{16}, \dfrac{42}{90}$

6. $\dfrac{13}{19}, \dfrac{26}{38}$

7. $\dfrac{3}{14}, \dfrac{21}{98}$

8. $\dfrac{12}{17}, \dfrac{50}{85}$

Solve each proportion. If necessary, round to the nearest hundredth.

9. $\dfrac{1}{a} = \dfrac{2}{14}$

10. $\dfrac{5}{b} = \dfrac{3}{9}$

11. $\dfrac{9}{g} = \dfrac{15}{10}$

12. $\dfrac{3}{a} = \dfrac{1}{6}$

13. $\dfrac{6}{z} = \dfrac{3}{5}$

14. $\dfrac{5}{f} = \dfrac{35}{21}$

15. $\dfrac{12}{7} = \dfrac{36}{m}$

16. $\dfrac{6}{23} = \dfrac{y}{69}$

17. $\dfrac{42}{56} = \dfrac{6}{f}$

18. $\dfrac{7}{b} = \dfrac{1}{9}$

19. $\dfrac{10}{14} = \dfrac{30}{m}$

20. $\dfrac{11}{15} = \dfrac{n}{60}$

21. $\dfrac{9}{c} = \dfrac{27}{39}$

22. $\dfrac{5}{12} = \dfrac{20}{g}$

23. $\dfrac{4}{21} = \dfrac{y}{84}$

24. $\dfrac{22}{x} = \dfrac{11}{30}$

25. **BOATING** Hue's boat used 5 gallons of gasoline in 4 hours. At this rate, how many gallons of gasoline will the boat use in 10 hours?

2-6 Practice

Ratios and Proportions

Determine whether each pair of ratios are equivalent ratios. Write *yes* or *no*.

1. $\dfrac{7}{6}, \dfrac{52}{48}$

2. $\dfrac{3}{11}, \dfrac{15}{66}$

3. $\dfrac{18}{24}, \dfrac{36}{48}$

4. $\dfrac{12}{11}, \dfrac{108}{99}$

5. $\dfrac{8}{9}, \dfrac{72}{81}$

6. $\dfrac{1.5}{9}, \dfrac{1}{6}$

7. $\dfrac{3.4}{5.2}, \dfrac{7.14}{10.92}$

8. $\dfrac{1.7}{1.2}, \dfrac{2.9}{2.4}$

9. $\dfrac{7.6}{1.8}, \dfrac{3.9}{0.9}$

Solve each proportion. If necessary, round to the nearest hundredth.

10. $\dfrac{5}{a} = \dfrac{30}{54}$

11. $\dfrac{v}{46} = \dfrac{34}{23}$

12. $\dfrac{40}{56} = \dfrac{k}{7}$

13. $\dfrac{28}{49} = \dfrac{4}{w}$

14. $\dfrac{3}{u} = \dfrac{27}{162}$

15. $\dfrac{y}{3} = \dfrac{48}{9}$

16. $\dfrac{2}{y} = \dfrac{10}{60}$

17. $\dfrac{5}{11} = \dfrac{35}{x}$

18. $\dfrac{3}{51} = \dfrac{z}{17}$

19. $\dfrac{6}{61} = \dfrac{12}{h}$

20. $\dfrac{g}{16} = \dfrac{6}{4}$

21. $\dfrac{14}{49} = \dfrac{2}{a}$

22. $\dfrac{7}{9} = \dfrac{8}{t}$

23. $\dfrac{3}{q} = \dfrac{5}{6}$

24. $\dfrac{m}{6} = \dfrac{5}{8}$

25. $\dfrac{v}{0.23} = \dfrac{7}{1.61}$

26. $\dfrac{3}{0.72} = \dfrac{12}{b}$

27. $\dfrac{6}{n} = \dfrac{3}{0.51}$

28. $\dfrac{7}{a-4} = \dfrac{14}{6}$

29. $\dfrac{3}{12} = \dfrac{2}{y+6}$

30. $\dfrac{m-1}{8} = \dfrac{2}{4}$

31. $\dfrac{5}{12} = \dfrac{x+1}{4}$

32. $\dfrac{r+2}{7} = \dfrac{5}{7}$

33. $\dfrac{3}{7} = \dfrac{x-2}{6}$

34. **PAINTING** Ysidra paints a room that has 400 square feet of wall space in $2\frac{1}{2}$ hours. At this rate, how long will it take her to paint a room that has 720 square feet of wall space?

35. **VACATION PLANS** Walker is planning a summer vacation. He wants to visit Petrified National Forest and Meteor Crater, Arizona, the 50,000-year-old impact site of a large meteor. On a map with a scale where 2 inches equals 75 miles, the two areas are about $1\frac{1}{2}$ inches apart. What is the distance between Petrified National Forest and Meteor Crater?

2-7 Skills Practice

Percent of Change

State whether each percent of change is a percent of _increase_ or a percent of _decrease_. Then find each percent of change. Round to the nearest whole percent.

1. original: 25
new: 10

2. original: 50
new: 75

3. original: 55
new: 50

4. original: 25
new: 28

5. original: 50
new: 30

6. original: 90
new: 95

7. original: 48
new: 60

8. original: 60
new: 45

Find the total price of each item.

9. dress: $69.00
tax: 5%

10. binder: $14.50
tax: 7%

11. hardcover book: $28.95
tax: 6%

12. groceries: $47.52
tax: 3%

13. filler paper: $6.00
tax: 6.5%

14. shoes: $65.00
tax: 4%

15. basketball: $17.00
tax: 6%

16. concert tickets: $48.00
tax: 7.5%

Find the discounted price of each item.

17. backpack: $56.25
discount: 20%

18. monitor: $150.00
discount: 50%

19. CD: $15.99
discount: 20%

20. shirt: $25.50
discount: 40%

21. sleeping bag: $125
discount: 25%

22. coffee maker: $102.00
discount: 45%

2-7 Practice

Percent of Change

State whether each percent of change is a percent of *increase* or a percent of *decrease*. Then find each percent of change. Round to the nearest whole percent.

1. original: 18
   new: 10

2. original: 140
   new: 160

3. original: 200
   new: 320

4. original: 10
   new: 25

5. original: 76
   new: 60

6. original: 128
   new: 120

7. original: 15
   new: 35.5

8. original: 98.6
   new: 64

9. original: 58.8
   new: 65.7

Find the total price of each item.

10. concrete blocks: $95.00
    tax: 6%

11. crib: $240.00
    tax: 6.5%

12. jacket: $125.00
    tax: 5.5%

13. class ring: $325.00
    tax: 6%

14. blanket: $24.99
    tax: 7%

15. kite: $18.90
    tax: 5%

Find the discounted price of each item.

16. dry cleaning: $25.00
    discount: 15%

17. computer game: $49.99
    discount: 25%

18. luggage: $185.00
    discount: 30%

19. stationery: $12.95
    discount: 10%

20. prescription glasses: $149
    discount: 20%

21. pair of shorts: $24.99
    discount: 45%

Find the final price of each item.

22. television: $375.00
    discount: 25%
    tax: 6%

23. DVD player: $269.00
    discount: 20%
    tax: 7%

24. printer: $255.00
    discount: 30%
    tax: 5.5%

25. **INVESTMENTS** The price per share of a stock decreased from $90 per share to $36 per share early in 2009. By what percent did the price of the stock decrease?

26. **HEATING COSTS** Customers of a utility company received notices in their monthly bills that heating costs for the average customer had increased 125% over last year because of an unusually severe winter. In January of last year, the Garcia's paid $120 for heating. What should they expect to pay this January if their bill increased by 125%?

2-8 Skills Practice

Literal Equations and Dimensional Analysis

Solve each equation or formula for the variable indicated.

1. $7t = x$, for t

2. $r = wp$, for p

3. $q - r = r$, for r

4. $4m - t = m$, for m

5. $7a - b = 15a$, for a

6. $-5c + d = 2c$, for c

7. $x - 2y = 1$, for y

8. $d + 3n = 1$, for n

9. $7f + g = 5$, for f

10. $ax - c = b$, for x

11. $rt - 2n = y$, for t

12. $bc + 3g = 2k$, for c

13. $kn + 4f = 9v$, for n

14. $8c + 6j = 5p$, for c

15. $\dfrac{x - c}{2} = d$, for x

16. $\dfrac{x - c}{2} = d$, for c

17. $\dfrac{p + 9}{5} = r$, for p

18. $\dfrac{b - 4z}{7} = a$, for b

19. The volume of a box V is given by the formula $V = \ell wh$, where ℓ is the length, w is the width, and h is the height.

    a. Solve the formula for h.

    b. What is the height of a box with a volume of 50 cubic meters, length of 10 meters, and width of 2 meters?

20. Trent purchases 44 euros worth of souvenirs while on vacation in France. If \$1 U.S. = 0.678 euros, find the cost of the souvenirs in United States dollars. Round to the nearest cent.

2-8 Practice

Literal Equations and Dimensional Analysis

Solve each equation or formula for the variable indicated.

1. $d = rt$, for r

2. $6w - y = 2z$, for w

3. $mx + 4y = 3t$, for x

4. $9s - 5g = -4u$, for s

5. $ab + 3c = 2x$, for b

6. $2p = kx - t$, for x

7. $\frac{2}{3}m + a = a + r$, for m

8. $\frac{2}{5}h + g = d$, for h

9. $\frac{2}{3}y + v = x$, for y

10. $\frac{3}{4}a - q = k$, for a

11. $\frac{rx + 9}{5} = h$, for x

12. $\frac{3b - 4}{2} = c$, for b

13. $2w - y = 7w - 2$, for w

14. $3\ell + y = 5 + 5\ell$, for ℓ

15. ELECTRICITY The formula for Ohm's Law is $E = IR$, where E represents voltage measured in volts, I represents current measured in amperes, and R represents resistance measured in ohms.

  a. Solve the formula for R.

  b. Suppose a current of 0.25 ampere flows through a resistor connected to a 12-volt battery. What is the resistance in the circuit?

16. MOTION In *uniform circular motion*, the speed v of a point on the edge of a spinning disk is $v = \frac{2\pi}{t}r$, where r is the radius of the disk and t is the time it takes the point to travel once around the circle.

  a. Solve the formula for r.

  b. Suppose a merry-go-round is spinning once every 3 seconds. If a point on the outside edge has a speed of 12.56 feet per second, what is the radius of the merry-go-round? (Use 3.14 for π.)

17. HIGHWAYS Interstate 90 is the longest interstate highway in the United States, connecting the cities of Seattle, Washington and Boston, Massachusetts. The interstate is 4,987,000 meters in length. If 1 mile = 1.609 kilometers, how many miles long is Interstate 90?

2-9 Skills Practice

Weighted Averages

1. **SEASONING** A health food store sells seasoning blends in bulk. One blend contains 20% basil. Sheila wants to add pure basil to some 20% blend to make 16 ounces of her own 30% blend. Let *b* represent the amount of basil Sheila should add to the 20% blend.

   a. Complete the table representing the problem.

	Ounces	Amount of Basil
20% Basil Blend		
100% Basil		
30% Basil Blend		

   b. Write an equation to represent the problem.

   c. How many ounces of basil should Sheila use to make the 30% blend?

   d. How many ounces of the 20% blend should she use?

2. **HIKING** At 7:00 A.M., two groups of hikers begin 21 miles apart and head toward each other. The first group, hiking at an average rate of 1.5 miles per hour, carries tents, sleeping bags, and cooking equipment. The second group, hiking at an average rate of 2 miles per hour, carries food and water. Let *t* represent the hiking time.

   a. Copy and complete the table representing the problem.

	r	*t*	*d = rt*
First group of hikers			
Second group of hikers			

   b. Write an equation using *t* that describes the distances traveled.

   c. How long will it be until the two groups of hikers meet?

3. **SALES** Sergio sells a mixture of Virginia peanuts and Spanish peanuts for $3.40 per pound. To make the mixture, he uses Virginia peanuts that cost $3.50 per pound and Spanish peanuts that cost $3.00 per pound. He mixes 10 pounds at a time.

   a. How many pounds of Virginia peanuts does Sergio use?

   b. How many pounds of Spanish peanuts does Sergio use?

2-9 Practice

Weighted Averages

1. GRASS SEED A nursery sells Kentucky Blue Grass seed for $5.75 per pound and Tall Fescue seed for $4.50 per pound. The nursery sells a mixture of the two kinds of seed for $5.25 per pound. Let k represent the amount of Kentucky Blue Grass seed the nursery uses in 5 pounds of the mixture.

a. Complete the table representing the problem.

	Number of Pounds	Price per Pound	Cost
Kentucky Blue Grass			
Tall Fescue			
Mixture			

b. Write an equation to represent the problem.

c. How much Kentucky Blue Grass does the nursery use in 5 pounds of the mixture?

d. How much Tall Fescue does the nursery use in 5 pounds of the mixture?

2. TRAVEL Two commuter trains carry passengers between two cities, one traveling east, and the other west, on different tracks. Their respective stations are 150 miles apart. Both trains leave at the same time, one traveling at an average speed of 55 miles per hour and the other at an average speed of 65 miles per hour. Let t represent the time until the trains pass each other.

a. Copy and complete the table representing the problem.

	r	t	$d = rt$
First Train			
Second Train			

b. Write an equation using t that describes the distances traveled.

c. How long after departing will the trains pass each other?

3. TRAVEL Two trains leave Raleigh at the same time, one traveling north, and the other south. The first train travels at 50 miles per hour and the second at 60 miles per hour. In how many hours will the trains be 275 miles apart?

4. JUICE A pineapple drink contains 15% pineapple juice. How much pure pineapple juice should be added to 8 quarts of the pineapple drink to obtain a mixture containing 50% pineapple juice?

3-1 Skills Practice

Graphing Linear Equations

Determine whether each equation is a linear equation. Write *yes* or *no*. If yes, write the equation in standard form.

1. $xy = 6$

2. $y = 2 - 3x$

3. $5x = y - 4$

4. $y = 2x + 5$

5. $y = -7 + 6x$

6. $y = 3x^2 + 1$

7. $y - 4 = 0$

8. $5x + 6y = 3x + 2$

9. $\frac{1}{2}y = 1$

Find the *x*- and *y*-intercepts of each linear function.

10.
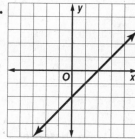

11.
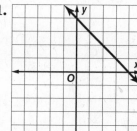

12.
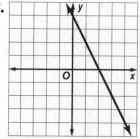

Graph each equation by making a table.

13. $y = 4$
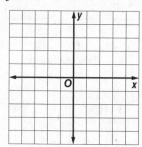

14. $y = 3x$
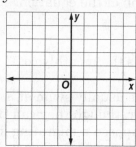

15. $y = x + 4$
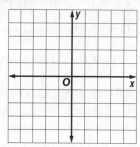

Graph each equation by using the *x*- and *y*-intercepts.

16. $x - y = 3$
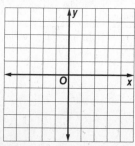

17. $10x = -5y$
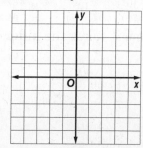

18. $4x = 2y + 6$
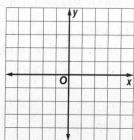

3-1 Practice

Graphing Linear Equations

Determine whether each equation is a linear equation. Write *yes* or *no*. If yes, write the equation in standard form and determine the *x*- and *y*-intercepts.

1. $4xy + 2y = 9$

2. $8x - 3y = 6 - 4x$

3. $7x + y + 3 = y$

4. $5 - 2y = 3x$

5. $\dfrac{x}{4} - \dfrac{y}{3} = 1$

6. $\dfrac{5}{x} - \dfrac{2}{y} = 7$

Graph each equation.

7. $\dfrac{1}{2}x - y = 2$

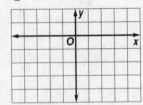

8. $5x - 2y = 7$

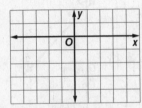

9. $1.5x + 3y = 9$

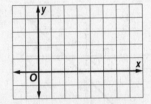

10. COMMUNICATIONS A telephone company charges $4.95 per month for long distance calls plus $0.05 per minute. The monthly cost c of long distance calls can be described by the equation $c = 0.05m + 4.95$, where m is the number of minutes.

a. Find the *y*-intercept of the graph of the equation.

b. Graph the equation.

c. If you talk 140 minutes, what is the monthly cost?

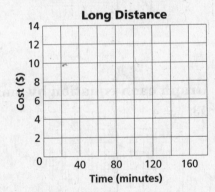

11. MARINE BIOLOGY Killer whales usually swim at a rate of 3.2–9.7 kilometers per hour, though they can travel up to 48.4 kilometers per hour. Suppose a migrating killer whale is swimming at an average rate of 4.5 kilometers per hour. The distance d the whale has traveled in t hours can be predicted by the equation $d = 4.5t$.

a. Graph the equation.

b. Use the graph to predict the time it takes the killer whale to travel 30 kilometers.

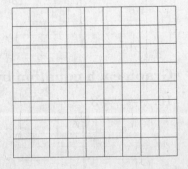

3-2 Skills Practice

Solving Linear Equations by Graphing

Solve each equation.

1. $2x - 5 = -3 + 2x$

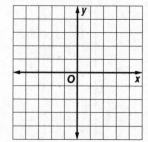

2. $-3x + 2 = 0$

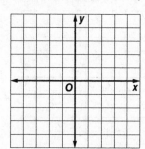

3. $3x + 2 = 3x - 1$

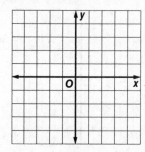

4. $4x - 1 = 4x + 2$

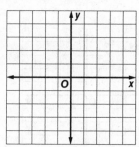

5. $4x - 1 = 0$

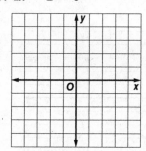

6. $0 = 5x + 3$

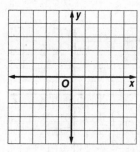

7. $0 = -2x + 4$

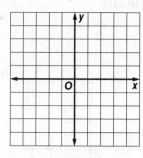

8. $-3x + 8 = 5 - 3x$

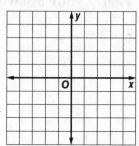

9. $-x + 1 = 0$

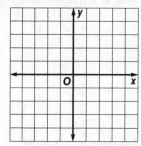

10. GIFT CARDS You receive a gift card for trading cards from a local store. The function $d = 20 - 1.95c$ represents the remaining dollars d on the gift card after obtaining c packages of cards. Find the zero of this function. Describe what this value means in this context.

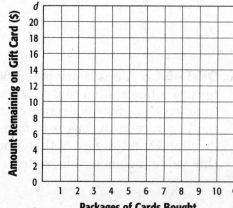

3-2 Practice

Solving Linear Equations by Graphing

Solve each equation.

1. $\frac{1}{2}x - 2 = 0$

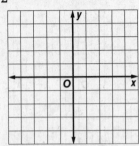

2. $-3x + 2 = -1$

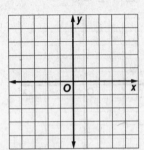

3. $4x - 2 = -2$

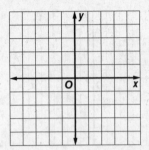

4. $\frac{1}{3}x + 2 = \frac{1}{3}x - 1$

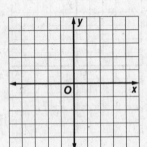

5. $\frac{2}{3}x + 4 = 3$

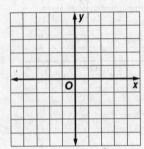

6. $\frac{3}{4}x + 1 = \frac{3}{4}x - 7$

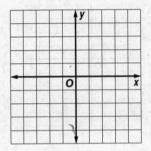

Solve each equation by graphing. Verify your answer algebraically

7. $13x + 2 = 11x - 1$

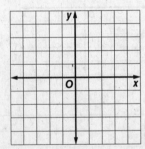

8. $-9x - 3 = -4x - 3$

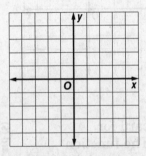

9. $-\frac{1}{3}x + 2 = \frac{2}{3}x - 1$

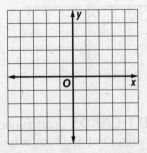

10. **DISTANCE** A bus is driving at 60 miles per hour toward a bus station that is 250 miles away. The function $d = 250 - 60t$ represents the distance d from the bus station the bus is t hours after it has started driving. Find the zero of this function. Describe what this value means in this context.

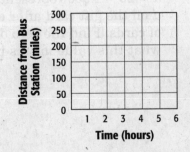

3-3 Skills Practice

Rate of Change and Slope

Find the slope of the line that passes through each pair of points.

1.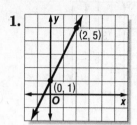

2.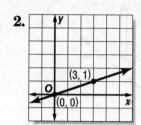

3.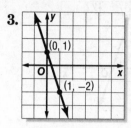

4. $(2, 5)$, $(3, 6)$

5. $(6, 1)$, $(-6, 1)$

6. $(4, 6)$, $(4, 8)$

7. $(5, 2)$, $(5, -2)$

8. $(2, 5)$, $(-3, -5)$

9. $(9, 8)$, $(7, -8)$

10. $(-5, -8)$, $(-8, 1)$

11. $(-3, 10)$, $(-3, 7)$

12. $(17, 18)$, $(18, 17)$

13. $(-6, -4)$, $(4, 1)$

14. $(10, 0)$, $(-2, 4)$

15. $(2, -1)$, $(-8, -2)$

16. $(5, -9)$, $(3, -2)$

17. $(12, 6)$, $(3, -5)$

18. $(-4, 5)$, $(-8, -5)$

19. $(-5, 6)$, $(7, -8)$

Find the value of r so the line that passes through each pair of points has the given slope.

20. $(r, 3)$, $(5, 9)$, $m = 2$

21. $(5, 9)$, $(r, -3)$, $m = -4$

22. $(r, 2)$, $(6, 3)$, $m = \frac{1}{2}$

23. $(r, 4)$, $(7, 1)$, $m =$

24. $(5, 3)$, $(r, -5)$, $m = 4$

25. $(7, r)$, $(4, 6)$, $m = 0$

3-3 Practice

Rate of Change and Slope

Find the slope of the line that passes through each pair of points.

1.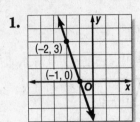

2.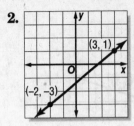

3.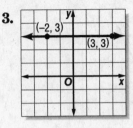

4. $(6, 3), (7, -4)$

5. $(-9, -3), (-7, -5)$

6. $(6, -2), (5, -4)$

7. $(7, -4), (4, 8)$

8. $(-7, 8), (-7, 5)$

9. $(5, 9), (3, 9)$

10. $(15, 2), (-6, 5)$

11. $(3, 9), (-2, 8)$

12. $(-2, -5), (7, 8)$

13. $(12, 10), (12, 5)$

14. $(0.2, -0.9), (0.5, -0.9)$

15. $\left(\frac{7}{3}, \frac{4}{3}\right), \left(-\frac{1}{3}, \frac{2}{3}\right)$

Find the value of *r* so the line that passes through each pair of points has the given slope.

16. $(-2, r), (6, 7), m = \frac{1}{2}$

17. $(-4, 3), (r, 5), m = \frac{1}{4}$

18. $(-3, -4), (-5, r), m = -\frac{9}{2}$

19. $(-5, r), (1, 3), m = \frac{7}{6}$

20. $(1, 4), (r, 5), m$ undefined

21. $(-7, 2), (-8, r), m = -5$

22. $(r, 7), (11, 8), m = -\frac{1}{5}$

23. $(r, 2), (5, r), m = 0$

24. ROOFING The *pitch* of a roof is the number of feet the roof rises for each 12 feet horizontally. If a roof has a pitch of 8, what is its slope expressed as a positive number?

25. SALES A daily newspaper had 12,125 subscribers when it began publication. Five years later it had 10,100 subscribers. What is the average yearly rate of change in the number of subscribers for the five-year period?

3-4 Skills Practice

Direct Variation

Name the constant of variation for each equation. Then determine the slope of the line that passes through each pair of points.

1.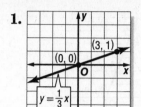

2.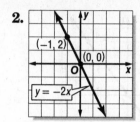

3.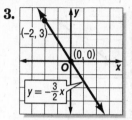

Graph each equation.

4. $y = 3x$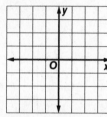

5. $y = -\frac{3}{4}x$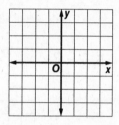

6. $y = \frac{2}{5}x$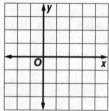

Suppose y varies directly as x. Write a direct variation equation that relates x and y. Then solve.

7. If $y = -8$ when $x = -2$, find x when $y = 32$.

8. If $y = 45$ when $x = 15$, find x when $y = 15$.

9. If $y = -4$ when x ___ 2, find y when $x = -6$.

10. If $y = -9$ when $x = 3$, find y when $x = -5$.

11. If $y = 4$ when $x = 16$, find y when $x = 6$.

12. If $y = 12$ when $x = 18$, find x when $y = -16$.

Write a direct variation equation that relates the variables. Then graph the equation.

13. **TRAVEL** The total cost C of gasoline is $3.00 times the number of gallons g.

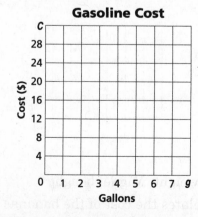

14. **SHIPPING** The number of delivered toys T is 3 times the total number of crates c.

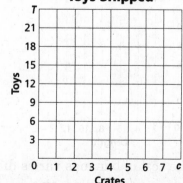

3-4 Practice

Direct Variation

Name the constant of variation for each equation. Then determine the slope of the line that passes through each pair of points.

1.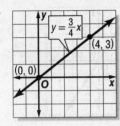

2.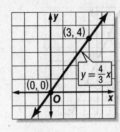

3.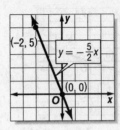

Graph each equation.

4. $y = -2x$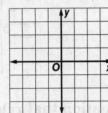

5. $y = \frac{6}{5}x$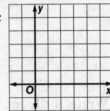

6. $y = -\frac{5}{2}x$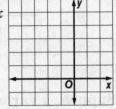

Suppose y varies directly as x. Write a direct variation equation that relates x and y. Then solve.

7. If $y = 7.5$ when $x = 0.5$, find y when $x = -0.3$.

8. If $y = 80$ when $x = 32$, find x when $y = 100$.

9. If $y = \frac{3}{4}$ when $x = 24$, find y when $x = 12$.

Write a direct variation equation that relates the variables. Then graph the equation.

10. **MEASURE** The width W of a rectangle is two thirds of the length ℓ.

11. **TICKETS** The total cost C of tickets is $4.50 times the number of tickets t.

Rectangle Dimensions

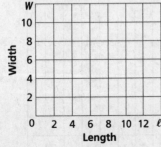

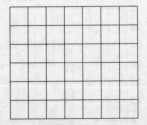

12. **PRODUCE** The cost of bananas varies directly with their weight. Miguel bought $3\frac{1}{2}$ pounds of bananas for $1.12. Write an equation that relates the cost of the bananas to their weight. Then find the cost of $4\frac{1}{4}$ pounds of bananas.

3-5 Skills Practice

Arithmetic Sequences as Linear Functions

Determine whether each sequence is an arithmetic sequence. Write *yes* or *no*. Explain.

1. 4, 7, 9, 12, . . .

2. 15, 13, 11, 9, . . .

3. 7, 10, 13, 16, . . .

4. −6, −5, −3, −1, . . .

5. −5, −3, −1, 1, . . .

6. −9, −12, −15, −18, . . .

7. 10, 15, 25, 40, . . .

8. −10, −5, 0, 5, . . .

Find the next three terms of each arithmetic sequence.

9. 3, 7, 11, 15, . . .

10. 22, 20, 18, 16, . . .

11. −13, −11, −9, −7 . . .

12. −2, −5, −8, −11, . . .

13. 19, 24, 29, 34, . . .

14. 16, 7, −2, −11, . . .

15. 2.5, 5, 7.5, 10, . . .

16. 3.1, 4.1, 5.1, 6.1, . . .

Write an equation for the *n*th term of each arithmetic sequence. Then graph the first five terms of the sequence.

17. 7, 13, 19, 25, . . .

18. 30, 26, 22, 18, . . .

19. −7, −4, −1, 2, . . .

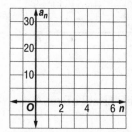

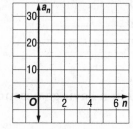

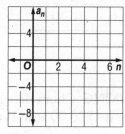

20. VIDEO DOWNLOADING Brian is downloading episodes of his favorite TV show to play on his personal media device. The cost to download 1 episode is $1.99. The cost to download 2 episodes is $3.98. The cost to download 3 episodes is $5.97. Write a function to represent the arithmetic sequence.

3-5 Practice

Arithmetic Sequences as Linear Functions

Determine whether each sequence is an arithmetic sequence. Write *yes* or *no*. Explain.

1. 21, 13, 5, −3, . . .

2. −5, 12, 29, 46, . . .

3. −2.2, −1.1, 0.1, 1.3, . . .

4. 1, 4, 9, 16, . . .

5. 9, 16, 23, 30, . . .

6. −1.2, 0.6, 1.8, 3.0, . . .

Find the next three terms of each arithmetic sequence.

7. 82, 76, 70, 64, . . .

8. −49, −35, −21, −7, . . .

9. $\frac{3}{4}, \frac{1}{2}, \frac{1}{4}, 0, . . .$

10. −10, −3, 4, 11 . . .

11. 12, 10, 8, 6, . . .

12. 12, 7, 2, −3, . . .

Write an equation for the *n*th term of each arithmetic sequence. Then graph the first five terms of the sequence.

13. 9, 13, 17, 21, . . .

14. −5, −2, 1, 4, . . .

15. 19, 31, 43, 55, . . .

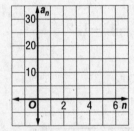

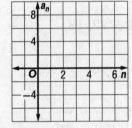

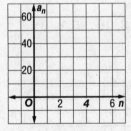

16. BANKING Chem deposited $115.00 in a savings account. Each week thereafter, he deposits $35.00 into the account.

   a. Write a function to represent the total amount Chem has deposited for any particular number of weeks after his initial deposit.

   b. How much has Chem deposited 30 weeks after his initial deposit?

17. STORE DISPLAYS Tamika is stacking boxes of tissue for a store display. Each row of tissues has 2 fewer boxes than the row below. The first row has 23 boxes of tissues.

   a. Write a function to represent the arithmetic sequence.

   b. How many boxes will there be in the tenth row?

3-6 Skills Practice

Proportional and Nonproportional Relationships

Write an equation in function notation for each relation.

1.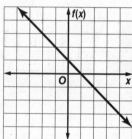

2.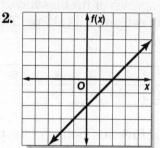

3.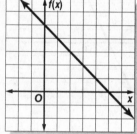

4.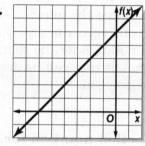

5.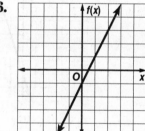

6.

7. **GAMESHOWS** The table shows how many points are awarded for answering consecutive questions on a gameshow.

Question answered	1	2	3	4	5
Points awarded	200	400	600	800	1000

a. Write an equation for the data given.

b. Find the number of points awarded if 9 questions were answered.

3-6 Practice

Proportional and Nonproportional Relationships

1. **BIOLOGY** Male fireflies flash in various patterns to signal location and perhaps to ward off predators. Different species of fireflies have different flash characteristics, such as the intensity of the flash, its rate, and its shape. The table below shows the rate at which a male firefly is flashing.

Times (seconds)	1	2	3	4	5
Number of Flashes	2	4	6	8	10

a. Write an equation in function notation for the relation.

b. How many times will the firefly flash in 20 seconds?

2. **GEOMETRY** The table shows the number of diagonals that can be drawn from one vertex in a polygon. Write an equation in function notation for the relation and find the number of diagonals that can be drawn from one vertex in a 12-sided polygon.

Sides	3	4	5	6
Diagonals	0	1	2	3

Write an equation in function notation for each relation.

3.

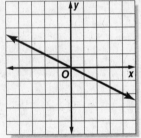

4.

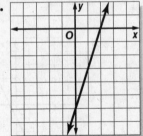

5.

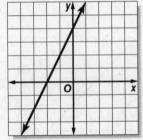

For each arithmetic sequence, determine the related function. Then determine if the function is *proportional* or *nonproportional*. Explain.

6. 1, 3, 5, . . .   7. 2, 7, 12, . . .   8. −3, −6, −9, . . .

4-1 Skills Practice

Graphing Equations in Slope-Intercept Form

Write an equation of a line in slope-intercept form with the given slope and *y*-intercept.

1. slope: 5, *y*-intercept: -3

2. slope: -2, *y*-intercept: 7

3. slope: -6, *y*-intercept: -2

4. slope: 7, *y*-intercept: 1

5. slope: 3, *y*-intercept: 2

6. slope: -4, *y*-intercept: -9

7. slope: 1, *y*-intercept: -12

8. slope: 0, *y*-intercept: 8

Write an equation in slope-intercept form for each graph shown.

9.
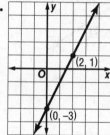

10.
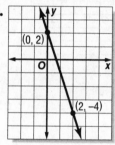

11.
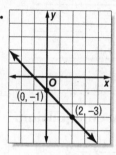

Graph each equation.

12. $y = x + 4$

13. $y = -2x - 1$

14. $x + y = -3$

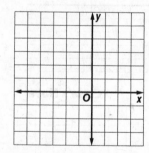

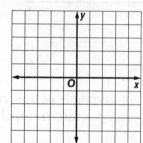

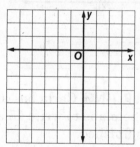

15. VIDEO RENTALS A video store charges $10 for a rental card plus $2 per rental.

a. Write an equation in slope-intercept form for the total cost *c* of buying a rental card and renting *m* movies.

b. Graph the equation.

c. Find the cost of buying a rental card and renting 6 movies.

Video Store Rental Costs

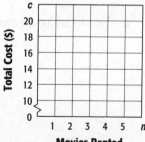

4-1 Practice

Graphing Equations in Slope-Intercept Form

Write an equation of a line in slope-intercept form with the given slope and *y*-intercept.

1. slope: $\frac{1}{4}$, *y*-intercept: 3

2. slope: $\frac{3}{2}$, *y*-intercept: −4

3. slope: 1.5, *y*-intercept: −1

4. slope: −2.5, *y*-intercept: 3.5

Write an equation in slope-intercept form for each graph shown.

5.
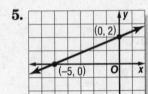

6.
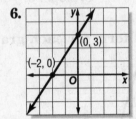

7.
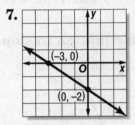

Graph each equation.

8. $y = -\frac{1}{2}x + 2$

9. $3y = 2x - 6$

10. $6x + 3y = 6$

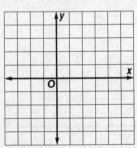

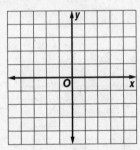

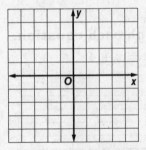

11. **WRITING** Carla has already written 10 pages of a novel. She plans to write 15 additional pages per month until she is finished.

   a. Write an equation to find the total number of pages *P* written after any number of months *m*.

   b. Graph the equation on the grid at the right.

   c. Find the total number of pages written after 5 months.

Carla's Novel

4-2 Skills Practice

Writing Equations in Slope-Intercept Form

Write an equation of the line that passes through the given point with the given slope.

1.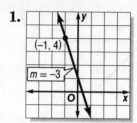

2.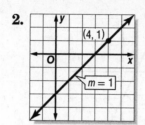

3.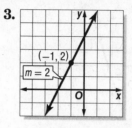

4. (1, 9); slope 4

5. (4, 2); slope −2

6. (2, −2); slope 3

7. (3, 0); slope 5

8. (−3, −2); slope 2

9. (−5, 4); slope −4

Write an equation of the line that passes through each pair of points.

10.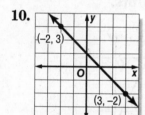

11.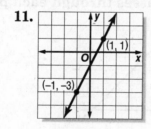

12.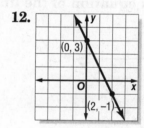

13. (1, 3), (−3, −5)

14. (1, 4), (6, −1)

15. (1, −1), (3, 5)

16. (−2, 4), (0, 6)

17. (3, 3), (1, −3)

18. (−1, 6), (3, −2)

19. INVESTING The price of a share of stock in XYZ Corporation was $74 two weeks ago. Seven weeks ago, the price was $59 a share.

    a. Write a linear equation to find the price p of a share of XYZ Corporation stock w weeks from now.

    b. Estimate the price of a share of stock five weeks ago.

4-2 Practice

Writing Equations in Slope-Intercept Form

Write an equation of the line that passes through the given point and has the given slope.

1.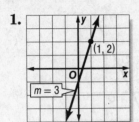

2.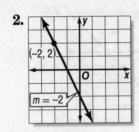

3.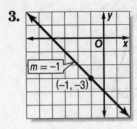

4. $(-5, 4)$; slope -3

5. $(4, 3)$; slope $\frac{1}{2}$

6. $(1, -5)$; slope $-\frac{3}{2}$

7. $(3, 7)$; slope $\frac{2}{7}$

8. $\left(-2, \frac{5}{2}\right)$; slope $-\frac{1}{2}$

9. $(5, 0)$; slope 0

Write an equation of the line that passes through each pair of points.

10.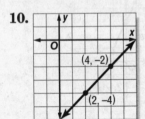

11.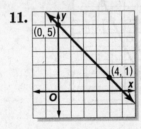

12.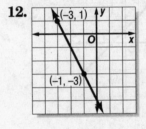

13. $(0, -4)$, $(5, -4)$

14. $(-4, -2)$, $(4, 0)$

15. $(-2, -3)$, $(4, 5)$

16. $(0, 1)$, $(5, 3)$

17. $(-3, 0)$, $(1, -6)$

18. $(1, 0)$, $(5, -1)$

19. DANCE LESSONS The cost for 7 dance lessons is $82. The cost for 11 lessons is $122. Write a linear equation to find the total cost C for ℓ lessons. Then use the equation to find the cost of 4 lessons.

20. WEATHER It is 76°F at the 6000-foot level of a mountain, and 49°F at the 12,000-foot level of the mountain. Write a linear equation to find the temperature T at an elevation x on the mountain, where x is in thousands of feet.

4-3 | **Skills Practice**

Writing Equations in Point-Slope Form

Write an equation in point-slope form for the line that passes through the given point with the slope provided.

1.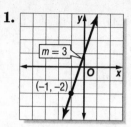

2.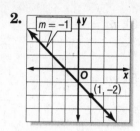

3.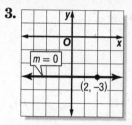

4. $(3, 1)$, $m = 0$

5. $(-4, 6)$, $m = 8$

6. $(1, -3)$, $m = -4$

7. $(4, -6)$, $m = 1$

8. $(3, 3)$, $m = \frac{4}{3}$

9. $(-5, -1)$, $m = -\frac{5}{4}$

Write each equation in standard form.

10. $y + 1 = x + 2$

11. $y + 9 = -3(x - 2)$

12. $y - 7 = 4(x + 4)$

13. $y - 4 = -(x - 1)$

14. $y - 6 = 4(x + 3)$

15. $y + 5 = -5(x - 3)$

16. $y - 10 = -2(x - 3)$

17. $y - 2 = -\frac{1}{2}(x - 4)$

18. $y + 11 = \frac{1}{3}(x + 3)$

Write each equation in slope-intercept form.

19. $y - 4 = 3(x - 2)$

20. $y + 2 = -(x + 4)$

21. $y - 6 = -2(x + 2)$

22. $y + 1 = -5(x - 3)$

23. $y - 3 = 6(x - 1)$

24. $y - 8 = 3(x + 5)$

25. $y - 2 = \frac{1}{2}(x + 6)$

26. $y + 1 = -\frac{1}{3}(x + 9)$

27. $y - \frac{1}{2} = x + \frac{1}{2}$

4-3   Practice

Writing Equations in Point-Slope Form

Write an equation in point-slope form for the line that passes through the given point with the slope provided.

1. $(2, 2)$, $m = -3$

2. $(1, -6)$, $m = -1$

3. $(-3, -4)$, $m = 0$

4. $(1, 3)$, $m = -\dfrac{3}{4}$

5. $(-8, 5)$, $m = -\dfrac{2}{5}$

6. $(3, -3)$, $m = \dfrac{1}{3}$

Write each equation in standard form.

7. $y - 11 = 3(x - 2)$

8. $y - 10 = -(x - 2)$

9. $y + 7 = 2(x + 5)$

10. $y - 5 = \dfrac{3}{2}(x + 4)$

11. $y + 2 = -\dfrac{3}{4}(x + 1)$

12. $y - 6 = \dfrac{4}{3}(x - 3)$

13. $y + 4 = 1.5(x + 2)$

14. $y - 3 = -2.4(x - 5)$

15. $y - 4 = 2.5(x + 3)$

Write each equation in slope-intercept form.

16. $y + 2 = 4(x + 2)$

17. $y + 1 = -7(x + 1)$

18. $y - 3 = -5(x + 12)$

19. $y - 5 = \dfrac{3}{2}(x + 4)$

20. $y - \dfrac{1}{4} = -3\left(x + \dfrac{1}{4}\right)$

21. $y - \dfrac{2}{3} = -2\left(x - \dfrac{1}{4}\right)$

22. CONSTRUCTION A construction company charges \$15 per hour for debris removal, plus a one-time fee for the use of a trash dumpster. The total fee for 9 hours of service is \$195.

   a. Write the point-slope form of an equation to find the total fee y for any number of hours x.

   b. Write the equation in slope-intercept form.

   c. What is the fee for the use of a trash dumpster?

23. MOVING There is a set daily fee for renting a moving truck, plus a charge of \$0.50 per mile driven. It costs \$64 to rent the truck on a day when it is driven 48 miles.

   a. Write the point-slope form of an equation to find the total charge y for any number of miles x for a one-day rental.

   b. Write the equation in slope-intercept form.

   c. What is the daily fee?

4-4   Skills Practice

Parallel and Perpendicular Lines

Write an equation in slope-intercept form for the line that passes through the given point and is parallel to the graph of each equation.

1.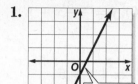

2.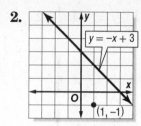

3.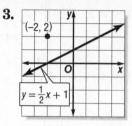

4. $(3, 2), y = 3x + 4$

5. $(-1, -2), y = -3x + 5$

6. $(-1, 1), y = x - 4$

7. $(1, -3), y = -4x - 1$

8. $(-4, 2), y = x + 3$

9. $(-4, 3), y = \frac{1}{2}x - 6$

10. **RADAR** On a radar screen, a plane located at $A(-2, 4)$ is flying toward $B(4, 3)$. Another plane, located at $C(-3, 1)$, is flying toward $D(3, 0)$. Are the planes' paths perpendicular? Explain.

Determine whether the graphs of the following equations are *parallel* or *perpendicular*. Explain.

11. $y = \frac{2}{3}x + 3, y = \frac{3}{2}x, 2x - 3y = 8$

12. $y = 4x, x + 4y = 12, 4x + y = 1$

Write an equation in slope-intercept form for the line that passes through the given point and is perpendicular to the graph of each equation.

13. $(-3, -2), y = x + 2$

14. $(4, -1), y = 2x - 4$

15. $(-1, -6), x + 3y = 6$

16. $(-4, 5), y = -4x - 1$

17. $(-2, 3), y = \frac{1}{4}x - 4$

18. $(0, 0), y = \frac{1}{2}x - 1$

4-4 Practice

Parallel and Perpendicular Lines

Write an equation in slope-intercept form for the line that passes through the given point and is parallel to the graph of each equation.

1. $(3, 2), y = x + 5$

2. $(-2, 5), y = -4x + 2$

3. $(4, -6), y = -\frac{3}{4}x + 1$

4. $(5, 4), y = \frac{2}{5}x - 2$

5. $(12, 3), y = \frac{4}{3}x + 5$

6. $(3, 1), 2x + y = 5$

7. $(-3, 4), 3y = 2x - 3$

8. $(-1, -2), 3x - y = 5$

9. $(-8, 2), 5x - 4y = 1$

10. $(-1, -4), 9x + 3y = 8$

11. $(-5, 6), 4x + 3y = 1$

12. $(3, 1), 2x + 5y = 7$

Write an equation in slope-intercept form for the line that passes through the given point and is perpendicular to the graph of each equation.

13. $(-2, -2), y = -\frac{1}{3}x + 9$

14. $(-6, 5), x - y = 5$

15. $(-4, -3), 4x + y = 7$

16. $(0, 1), x + 5y = 15$

17. $(2, 4), x - 6y = 2$

18. $(-1, -7), 3x + 12y = -6$

19. $(-4, 1), 4x + 7y = 6$

20. $(10, 5), 5x + 4y = 8$

21. $(4, -5), 2x - 5y = -10$

22. $(1, 1), 3x + 2y = -7$

23. $(-6, -5), 4x + 3y = -6$

24. $(-3, 5), 5x - 6y = 9$

25. GEOMETRY Quadrilateral $ABCD$ has diagonals $\overline{AC}$ and $\overline{BD}$. Determine whether $\overline{AC}$ is perpendicular to $\overline{BD}$. Explain.

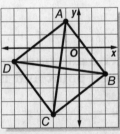

26. GEOMETRY Triangle ABC has vertices $A(0, 4)$, $B(1, 2)$, and $C(4, 6)$. Determine whether triangle ABC is a right triangle. Explain.

4-5 | **Skills Practice**

Scatter Plots and Lines of Fit

Determine whether each graph shows a *positive correlation*, a *negative correlation*, or *no correlation*. If there is a positive or negative correlation, describe its meaning in the situation.

1.
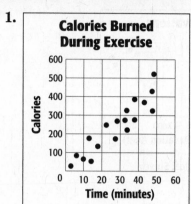

2.
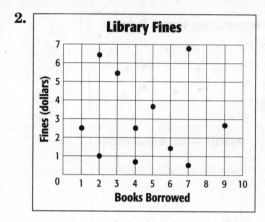

3.
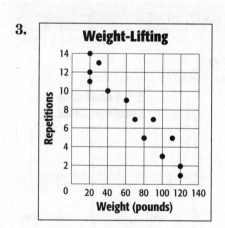

4.
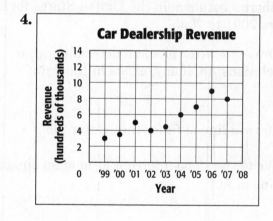

5. **BASEBALL** The scatter plot shows the average price of a major-league baseball ticket from 1997 to 2006.

a. Determine what relationship, if any, exists in the data. Explain.

b. Use the points (1998, 13.60) and (2003, 19.00) to write the slope-intercept form of an equation for the line of fit shown in the scatter plot.

c. Predict the price of a ticket in 2009.

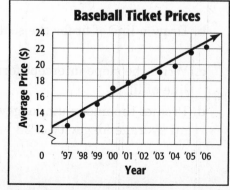

Source: Team Marketing Report, Chicago

4-5 Practice

Scatter Plots and Lines of Fit

Determine whether each graph shows a *positive correlation*, a *negative correlation*, or *no correlation*. If there is a positive or negative correlation, describe its meaning in the situation.

1.
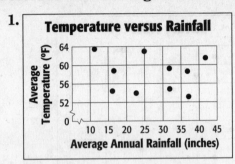
Source: National Oceanic and Atmospheric Administration

2.
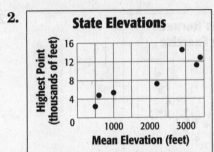
Source: U.S. Geological Survey

3. **DISEASE** The table shows the number of cases of foodborne botulism in the United States for the years 2001 to 2005.

   a. Draw a scatter plot and determine what relationship, if any, exists in the data.

   b. Draw a line of fit for the scatter plot.

   c. Write the slope-intercept form of an equation for the line of fit.

U.S. Foodborne Botulism Cases				
Year	2001	2002	2003	2004	2005
Cases	39	28	20	16	18

Source: Centers for Disease Control

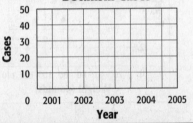

4. **ZOOS** The table shows the average and maximum longevity of various animals in captivity.

   a. Draw a scatter plot and determine what relationship, if any, exists in the data.

   b. Draw a line of fit for the scatter plot.

   c. Write the slope-intercept form of an equation for the line of fit.

   d. Predict the maximum longevity for an animal with an average longevity of 33 years.

Longevity (years)							
Avg.	12	25	15	8	35	40	41	20
Max.	47	50	40	20	70	77	61	54

Source: Walker's Mammals of the World

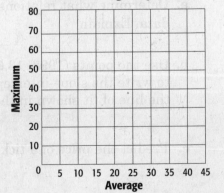

4-6 Skills Practice

Regression and Median-Fit Lines

Write an equation of the regression line for the data in each table below. Then find the correlation coefficient.

1. **SOCCER** The table shows the number of goals a soccer team scored each season since 2002.

Year	2002	2003	2004	2005	2006	2007
Goals Scored	42	48	46	50	52	48

2. **PHYSICAL FITNESS** The table shows the percentage of seventh grade students in public school who met all six of California's physical fitness standards each year since 2002.

Year	2002	2003	2004	2005	2006
Percentage	24.0%	36.4%	38.0%	40.8%	37.5%

Source: California Department of Education

3. **TAXES** The table shows the estimated sales tax revenues, in billions of dollars, for Massachusetts each year since 2004.

Year	2004	2005	2006	2007	2008
Tax Revenue	3.75	3.89	4.00	4.17	4.47

Source: Beacon Hill Institute

4. **PURCHASING** The SureSave supermarket chain closely monitors how many diapers are sold each year so that they can reasonably predict how many diapers will be sold in the following year.

Year	2003	2004	2005	2006	2007
Diapers Sold	60,200	65,000	66,300	65,200	70,600

   a. Find an equation for the median-fit line.

   b. How many diapers should SureSave anticipate selling in 2008?

5. **FARMING** Some crops, such as barley, are very sensitive to how acidic the soil is. To determine the ideal level of acidity, a farmer measured how many bushels of barley he harvests in different fields with varying acidity levels.

Soil Acidity (pH)	5.7	6.2	6.6	6.8	7.1
Bushels Harvested	3	20	48	61	73

   a. Find an equation for the regression line.

   b. According to the equation, how many bushels would the farmer harvest if the soil had a pH of 10?

   c. Is this a reasonable prediction? Explain.

4-6 Practice

Regression and Median-Fit Lines

Write an equation of the regression line for the data in each table below. Then find the correlation coefficient.

1. TURTLES The table shows the number of turtles hatched at a zoo each year since 2002.

Year	2003	2004	2005	2006	2007
Turtles Hatched	21	17	16	16	14

2. SCHOOL LUNCHES The table shows the percentage of students receiving free or reduced price school lunches in Marin County, California each year since 2003.

Year	2003	2004	2005	2006	2007
Percentage	14.4%	15.8%	18.3%	18.6%	20.9%

Source: KidsData

3. SPORTS Below is a table showing the number of students signed up to play lacrosse after school in each age group.

Age	13	14	15	16	17
Lacrosse Players	17	14	6	9	12

4. LANGUAGE The State of California keeps track of how many millions of students are learning English as a second language each year.

Year	2003	2004	2005	2006	2007
English Learners	1.600	1.599	1.592	1.570	1.569

Source: California Department of Education

a. Find an equation for the median-fit line.

b. Predict the number of students who were learning English in California in 2001.

c. Predict the number of students who will be learning English in California in 2010.

5. POPULATION Detroit, Michigan, like a number of large cities, is losing population every year. Below is a table showing the population of Detroit each decade.

Year	1960	1970	1980	1990	2000
Population (millions)	1.67	1.51	1.20	1.03	0.95

Source: U.S. Census Bureau

a. Find an equation for the regression line.

b. Find the correlation coefficient and explain the meaning of its sign.

c. Estimate the population of Detroit in 2008.

4-7 Skills Practice

Special Functions

Graph each function. State the domain and range.

1. $f(x) = [\![x - 2]\!]$

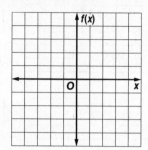

2. $f(x) = 3[\![x]\!]$

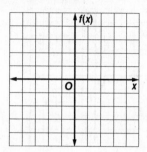

3. $f(x) = [\![2x]\!]$

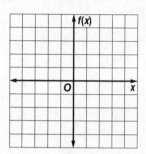

4. $f(x) = |x| - 3$

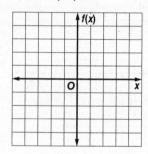

5. $f(x) = |2x|$

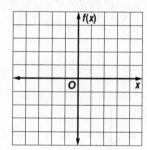

6. $f(x) = |2x + 5|$

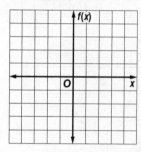

7. $f(x) = \begin{cases} 2x & \text{if } x \leq 1 \\ -x + 3 & \text{if } x > 2 \end{cases}$

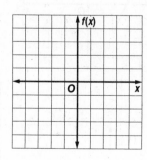

8. $f(x) = \begin{cases} x + 4 & \text{if } x \leq 1 \\ 0.25x + 1 & \text{if } x > 1 \end{cases}$

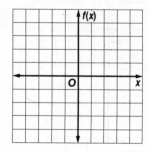

9. $f(x) = \begin{cases} x + 2 & \text{if } x < 0 \\ -0.5x + 1 & \text{if } x \geq 0 \end{cases}$

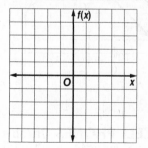

4-7 Practice

Special Functions

Graph each function. State the domain and range.

1. $f(x) = -2[\![x + 1]\!]$

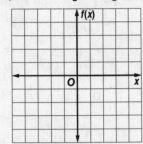

2. $f(x) = [\![x + 3]\!] - 2$

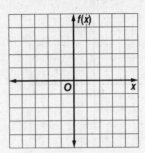

3. $f(x) = -\left|\frac{1}{2}x\right| + 1$

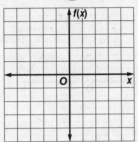

4. $f(x) = |2x + 4| - 3$

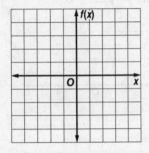

5. $f(x) = \begin{cases} 2 & \text{if } x > -1 \\ x + 4 & \text{if } x \le -1 \end{cases}$

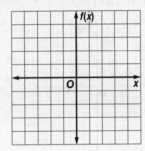

6. $f(x) = \begin{cases} -2x + 3 & \text{if } x > 0 \\ \frac{1}{2}x - 1 & \text{if } x \le 0 \end{cases}$

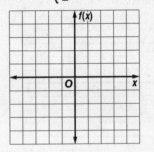

Determine the domain and range of each function.

7.

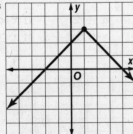

8.

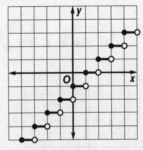

9.

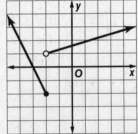

10. CELL PHONES Jacob's cell phone service costs $5 each month plus $0.35 for each minute he uses. Every fraction of a minute is rounded up to the next minute.

a. Draw a graph to represent the cost of using the cell phone.

b. What is Jacob's monthly bill if he uses 124.8 minutes?

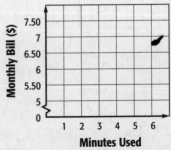

5-1 Skills Practice

Solving Inequalities by Addition and Subtraction

Match each inequality to the graph of its solution.

1. $x + 11 > 16$

a. $-8\ -7\ -6\ -5\ -4\ -3\ -2\ -1\ 0$

2. $x - 6 < 1$

b. $-4\ -3\ -2\ -1\ 0\ 1\ 2\ 3\ 4$

3. $x + 2 \le -3$

c. $0\ 1\ 2\ 3\ 4\ 5\ 6\ 7\ 8$

4. $x + 3 \ge 1$

d. $-8\ -7\ -6\ -5\ -4\ -3\ -2\ -1\ 0$

5. $x - 1 < -7$

e. $0\ 1\ 2\ 3\ 4\ 5\ 6\ 7\ 8$

Solve each inequality. Check your solution, and then graph it on a number line.

6. $d - 5 \le 1$

$0\ 1\ 2\ 3\ 4\ 5\ 6\ 7\ 8$

7. $t + 9 < 8$

$-4\ -3\ -2\ -1\ 0\ 1\ 2\ 3\ 4$

8. $a - 7 > -13$

$-8\ -7\ -6\ -5\ -4\ -3\ -2\ -1\ 0$

9. $w - 1 < 4$

$0\ 1\ 2\ 3\ 4\ 5\ 6\ 7\ 8$

10. $4 \ge k + 3$

$-4\ -3\ -2\ -1\ 0\ 1\ 2\ 3\ 4$

11. $-9 \le b - 4$

$-8\ -7\ -6\ -5\ -4\ -3\ -2\ -1\ 0$

12. $-2 \ge x + 4$

$-8\ -7\ -6\ -5\ -4\ -3\ -2\ -1\ 0$

13. $2y < y + 2$

$-4\ -3\ -2\ -1\ 0\ 1\ 2\ 3\ 4$

Define a variable, write an inequality, and solve each problem. Check your solution.

14. A number decreased by 10 is greater than -5.

15. A number increased by 1 is less than 9.

16. Seven more than a number is less than or equal to -18.

17. Twenty less than a number is at least 15.

18. A number plus 2 is at most 1.

5-1 Practice

Solving Inequalities by Addition and Subtraction

Match each inequality with its corresponding graph.

1. $-8 \geq x - 15$

a.

2. $4x + 3 < 5x$

b.

3. $8x > 7x - 4$

c.

4. $12 + x \leq 9$

d.

Solve each inequality. Check your solution, and then graph it on a number line.

5. $r - (-5) > -2$

6. $3x + 8 \geq 4x$

7. $n - 2.5 \geq -5$

8. $1.5 < y + 1$

9. $z + 3 > \dfrac{2}{3}$

10. $\dfrac{1}{2} \leq c - \dfrac{3}{4}$

Define a variable, write an inequality, and solve each problem. Check your solution.

11. The sum of a number and 17 is no less than 26.

12. Twice a number minus 4 is less than three times the number.

13. Twelve is at most a number decreased by 7.

14. Eight plus four times a number is greater than five times the number.

15. ATMOSPHERIC SCIENCE The troposphere extends from the Earth's surface to a height of 6–12 miles, depending on the location and the season. If a plane is flying at an altitude of 5.8 miles, and the troposphere is 8.6 miles deep in that area, how much higher can the plane go without leaving the troposphere?

16. EARTH SCIENCE Mature soil is composed of three layers, the uppermost being topsoil. Jamal is planting a bush that needs a hole 18 centimeters deep for the roots. The instructions suggest an additional 8 centimeters depth for a cushion. If Jamal wants to add even more cushion, and the topsoil in his yard is 30 centimeters deep, how much more cushion can he add and still remain in the topsoil layer?

5-2 Skills Practice

Solving Inequalities by Multiplication and Division

Match each inequality with its corresponding statement.

1. $3n < 9$ **a.** Three times a number is at most nine.

2. $\frac{1}{3}n \geq 9$ **b.** One third of a number is no more than nine.

3. $3n \leq 9$ **c.** Negative three times a number is more than nine.

4. $-3n > 9$ **d.** Three times a number is less than nine.

5. $\frac{1}{3}n \leq 9$ **e.** Negative three times a number is at least nine.

6. $-3n \geq 9$ **f.** One third of a number is greater than or equal to nine.

Solve each inequality. Check your solution.

7. $14g > 56$ **8.** $11w \leq 77$ **9.** $20b \geq -120$ **10.** $-8r < 16$

11. $-15p \leq -90$ **12.** $\frac{x}{4} < 9$ **13.** $\frac{a}{9} \geq -15$ **14.** $-\frac{p}{7} > -9$

15. $-\frac{t}{12} \geq 6$ **16.** $5z < -90$ **17.** $-13m > -26$ **18.** $\frac{k}{5} \leq -17$

19. $-y < 36$ **20.** $-16c \geq -224$ **21.** $-\frac{h}{10} \leq 2$ **22.** $12 > \frac{d}{12}$

Define a variable, write an inequality, and solve each problem. Check your solution.

23. Four times a number is greater than -48.

24. One eighth of a number is less than or equal to 3.

25. Negative twelve times a number is no more than 84.

26. Negative one sixth of a number is less than -9.

27. Eight times a number is at least 16.

5-2 Practice

Solving Inequalities by Multiplication and Division

Match each inequality with its corresponding statement.

1. $-4n \geq 5$ **a.** Negative four times a number is less than five.

2. $\frac{4}{5}n > 5$ **b.** Four fifths of a number is no more than five.

3. $4n \leq 5$ **c.** Four times a number is fewer than five.

4. $\frac{4}{5}n \leq 5$ **d.** Negative four times a number is no less than five.

5. $4n < 5$ **e.** Four times a number is at most five.

6. $-4n < 5$ **f.** Four fifths of a number is more than five.

Solve each inequality. Check your solution.

7. $-\frac{a}{5} < -14$ **8.** $-13h \leq 52$ **9.** $\frac{b}{16} \geq -6$ **10.** $39 > 13p$

11. $\frac{2}{3}n > -12$ **12.** $-\frac{5}{9}t < 25$ **13.** $-\frac{3}{5}m \leq -6$ **14.** $\frac{10}{3}k \geq -10$

15. $-3b \leq 0.75$ **16.** $-0.9c > -9$ **17.** $0.1x \geq -4$ **18.** $-2.3 < \frac{j}{4}$

19. $-15y < 3$ **20.** $2.6v \geq -20.8$ **21.** $0 > -0.5u$ **22.** $\frac{7}{8}f \leq -1$

Define a variable, write an inequality, and solve each problem. Check your solution.

23. Negative three times a number is at least 57.

24. Two thirds of a number is no more than -10.

25. Negative three fifths of a number is less than -6.

26. FLOODING A river is rising at a rate of 3 inches per hour. If the river rises more than 2 feet, it will exceed flood stage. How long can the river rise at this rate without exceeding flood stage?

27. SALES Pet Supplies makes a profit of $5.50 per bag on its line of natural dog food. If the store wants to make a profit of no less than $5225 on natural dog food, how many bags of dog food does it need to sell?

5-3 Skills Practice
Solving Multi-Step Inequalities

Justify each indicated step.

1.　　$\frac{3}{4}t - 3 \geq -15$

$\frac{3}{4}t - 3 + 3 \geq -15 + 3$　　**a.** ___?___

$\frac{3}{4}t \geq -12$

$\frac{4}{3}\left(\frac{3}{4}\right)t \geq \frac{4}{3}(-12)$　　**b.** ___?___

$t \geq -16$

a. Add 3 to each side.
b. Multiply each side by $\frac{4}{3}$.

2. $5(k + 8) - 7 \leq 23$

$5k + 40 - 7 \leq 23$　　**a.** ___?___

$5k + 33 \leq 23$

$5k + 33 - 33 \leq 23 - 33$　　**b.** ___?___

$5k \leq -10$

$\frac{5k}{5} \leq \frac{-10}{5}$　　**c.** ___?___

$k \leq -2$

a. Distributive Property
b. Subtract 33 from each side.
c. Divide each side by 5.

Solve each inequality. Check your solution.

3. $-2b + 4 > -6$

4. $3x + 15 \leq 21$

5. $\frac{d}{2} - 1 \geq 3$

6. $\frac{2}{5}a - 4 < 2$

7. $-\frac{t}{5} + 7 > -4$

8. $\frac{3}{4}j - 10 \geq 5$

9. $-\frac{2}{3}f + 3 < -9$

10. $2p + 5 \geq 3p - 10$

11. $4k + 15 > -2k + 3$

12. $2(-3m - 5) \geq -28$

13. $-6(w + 1) < 2(w + 5)$

14. $2(q - 3) + 6 \leq -10$

Define a variable, write an inequality, and solve each problem.
Check your solution.

15. Four more than the quotient of a number and three is at least nine.

16. The sum of a number and fourteen is less than or equal to three times the number.

17. Negative three times a number increased by seven is less than negative eleven.

18. Five times a number decreased by eight is at most ten more than twice the number.

19. Seven more than five sixths of a number is more than negative three.

20. Four times the sum of a number and two increased by three is at least twenty-seven.

5-3 Practice

Solving Multi-Step Inequalities

Justify each indicated step.

1. $x > \dfrac{5x - 12}{8}$

    $8x > (8)\dfrac{5x - 12}{8}$ **a.** ?

    $8x > 5x - 12$

    $8x - 5x > 5x - 12 - 5x$ **b.** ?

    $3x > -12$

    $\dfrac{3x}{3} > \dfrac{-12}{3}$ **c.** ?

    $x > -4$

2. $2(2h + 2) < 2(3h + 5) - 12$

    $4h + 4 < 6h + 10 - 12$ **a.** ?

    $4h + 4 < 6h - 2$

    $4h + 4 - 6h < 6h - 2 - 6h$ **b.** ?

    $-2h + 4 < -2$

    $-2h + 4 - 4 < -2 - 4$ **c.** ?

    $-2h < -6$

    $\dfrac{-2h}{-2} > \dfrac{-6}{-2}$ **d.** ?

    $h > 3$

Solve each inequality. Check your solution.

3. $-5 - \dfrac{t}{6} \geq -9$

4. $4u - 6 \geq 6u - 20$

5. $13 > \dfrac{2}{3}a - 1$

6. $\dfrac{w + 3}{2} < -8$

7. $\dfrac{3f - 10}{5} > 7$

8. $h \leq \dfrac{6h + 3}{5}$

9. $3(z + 1) + 11 < -2(z + 13)$

10. $3r + 2(4r + 2) \leq 2(6r + 1)$

11. $5n - 3(n - 6) \geq 0$

Define a variable, write an inequality, and solve each problem. Check your solution.

12. A number is less than one fourth the sum of three times the number and four.

13. Two times the sum of a number and four is no more than three times the sum of the number and seven decreased by four.

14. **GEOMETRY** The area of a triangular garden can be no more than 120 square feet. The base of the triangle is 16 feet. What is the height of the triangle?

15. **MUSIC PRACTICE** Nabuko practices the violin at least 12 hours per week. She practices for three fourths of an hour each session. If Nabuko has already practiced 3 hours in one week, how many sessions remain to meet or exceed her weekly practice goal?

NAME _____ DATE _____ PERIOD _____

5-4 Skills Practice

Solving Compound Inequalities

Graph the solution set of each compound inequality.

1. $b > 3$ or $b \le 0$

-4 -3 -2 -1 0 1 2 3 4

2. $z \le 3$ and $z \ge -2$

-4 -3 -2 -1 0 1 2 3 4

3. $k > 1$ and $k > 5$

0 1 2 3 4 5 6 7 8

4. $y < -1$ or $y \ge 1$

-4 -3 -2 -1 0 1 2 3 4

Write a compound inequality for each graph.

5.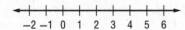
-4 -3 -2 -1 0 1 2 3 4

6.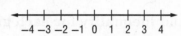
-2 -1 0 1 2 3 4 5 6

7.
-4 -3 -2 -1 0 1 2 3 4

8.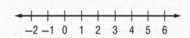
-4 -3 -2 -1 0 1 2 3 4

Solve each compound inequality. Then graph the solution set.

9. $m + 3 \ge 5$ and $m + 3 < 7$

-2 -1 0 1 2 3 4 5 6

10. $y - 5 < -4$ or $y - 5 \ge 1$

-2 -1 0 1 2 3 4 5 6

11. $4 < f + 6$ and $f + 6 < 5$

-4 -3 -2 -1 0 1 2 3 4

12. $w + 3 \le 0$ or $w + 7 \ge 9$

-4 -3 -2 -1 0 1 2 3 4

13. $-6 < b - 4 < 2$

-2 -1 0 1 2 3 4 5 6

14. $p - 2 \le -2$ or $p - 2 > 1$

-4 -3 -2 -1 0 1 2 3 4

Define a variable, write an inequality, and solve each problem. Check your solution.

15. A number plus one is greater than negative five and less than three.

16. A number decreased by two is at most four or at least nine.

17. The sum of a number and three is no more than eight or is more than twelve.

5-4   Practice

Solving Compound Inequalities

Graph the solution set of each compound inequality.

1. $-4 \le n \le 1$ ⟨─┼─┼─┼─┼─┼─┼─┼─┼─┼─┼─⟩
-6 -5 -4 -3 -2 -1 0 1 2

2. $x > 0$ or $x < 3$ ⟨─┼─┼─┼─┼─┼─┼─┼─┼─┼─⟩
-4 -3 -2 -1 0 1 2 3 4

3. $g < -3$ or $g \ge 4$ ⟨─┼─┼─┼─┼─┼─┼─┼─┼─┼─┼─⟩

4. $-4 \le p \le 4$ ⟨─┼─┼─┼─┼─┼─┼─┼─┼─┼─┼─⟩

Write a compound inequality for each graph.

5.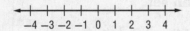
-4 -3 -2 -1 0 1 2 3 4

6.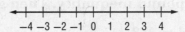
-2 -1 0 1 2 3 4 5 6

7.
-2 -1 0 1 2 3 4 5 6

8.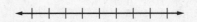
-6 -5 -4 -3 -2 -1 0 1 2

Solve each compound inequality. Then graph the solution set.

9. $k - 3 < -7$ or $k + 5 \ge 8$

⟨─┼─┼─┼─┼─┼─┼─┼─┼─┼─⟩
-4 -3 -2 -1 0 1 2 3 4

10. $-n < 2$ or $2n - 3 > 5$

⟨─┼─┼─┼─┼─┼─┼─┼─┼─┼─⟩
-4 -3 -2 -1 0 1 2 3 4

11. $5 < 3h + 2 \le 11$

⟨─┼─┼─┼─┼─┼─┼─┼─┼─┼─⟩

12. $2c - 4 > -6$ and $3c + 1 < 13$

⟨─┼─┼─┼─┼─┼─┼─┼─┼─┼─⟩

Define a variable, write an inequality, and solve each problem. Check your solution.

13. Two times a number plus one is greater than five and less than seven.

14. A number minus one is at most nine, or two times the number is at least twenty-four.

15. METEOROLOGY Strong winds called the prevailing westerlies blow from west to east in a belt from 40° to 60° latitude in both the Northern and Southern Hemispheres.

a. Write an inequality to represent the latitude of the prevailing westerlies.

b. Write an inequality to represent the latitudes where the prevailing westerlies are *not* located.

16. NUTRITION A cookie contains 9 grams of fat. If you eat no fewer than 4 and no more than 7 cookies, how many grams of fat will you consume?

5-5 Skills Practice

Inequalities Involving Absolute Value

Match each open sentence with the graph of its solution set.

1. $|x| > 2$

a.
$$-5\ -4\ -3\ -2\ -1\ \ 0\ \ 1\ \ 2\ \ 3\ \ 4\ \ 5$$

2. $|x - 2| \le 3$

b.
$$-5\ -4\ -3\ -2\ -1\ \ 0\ \ 1\ \ 2\ \ 3\ \ 4\ \ 5$$

3. $|x + 1| < 4$

c.
$$-4\ -3\ -2\ -1\ \ 0\ \ 1\ \ 2\ \ 3\ \ 4\ \ 5\ \ 6$$

Express each statement using an inequality involving absolute value.

4. The weatherman predicted that the temperature would be within 3° of 52°F.

5. Serena will make the B team if she scores within 8 points of the team average of 92.

6. The dance committee expects attendance to number within 25 of last year's 87 students.

Solve each inequality. Then graph the solution set.

7. $|x + 1| < 0$
$$-6\ -5\ -4\ -3\ -2\ -1\ \ 0\ \ 1\ \ 2\ \ 3\ \ 4$$

8. $|c - 3| < 1$
$$-3\ -2\ -1\ \ 0\ \ 1\ \ 2\ \ 3\ \ 4\ \ 5\ \ 6\ \ 7$$

9. $|n + 2| \ge 1$
$$-6\ -5\ -4\ -3\ -2\ -1\ \ 0\ \ 1\ \ 2\ \ 3\ \ 4$$

10. $|t + 6| > 4$
$$-10\ -9\ -8\ -7\ -6\ -5\ -4\ -3\ -2\ -1\ \ 0$$

11. $|w - 2| < 2$
$$-4\ -3\ -2\ -1\ \ 0\ \ 1\ \ 2\ \ 3\ \ 4\ \ 5\ \ 6$$

12. $|k - 5| \le 4$
$$0\ \ 1\ \ 2\ \ 3\ \ 4\ \ 5\ \ 6\ \ 7\ \ 8\ \ 9\ \ 10$$

5-5 Practice

Inequalities Involving Absolute Value

Match each open sentence with the graph of its solution set.

1. $|x - 3| \geq 1$

a.

2. $|2x + 1| < 5$

b.

3. $|5 - x| \geq 3$

c.

Express each statement using an inequality involving absolute value.

4. The height of the plant must be within 2 inches of the standard 13-inch show size.

5. The majority of grades in Sean's English class are within 4 points of 85.

Solve each inequality. Then graph the solution set.

6. $|2z - 9| \leq 1$

7. $|3 - 2r| > 7$

8. $|3t + 6| < 9$

9. $|2g - 5| \geq 9$

Write an open sentence involving absolute value for each graph.

10.

11.

12.

13.

14. RESTAURANTS The menu at Jeanne's favorite restaurant states that the roasted chicken with vegetables entree typically contains 480 Calories. Based on the size of the chicken, the actual number of Calories in the entree can vary by as many as 40 Calories from this amount.

a. Write an absolute value inequality to represent the situation.

b. What is the range of the number of Calories in the chicken entree?

5-6 Skills Practice

Graphing Inequalities in Two Variables

Match each inequality to the graph of its solution.

1. $y - 2x < 2$

2. $y \leq -3x$

3. $2y - x \geq 4$

4. $x + y > 1$

a.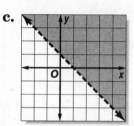

b.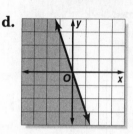

c.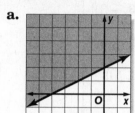

d.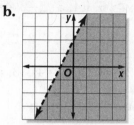

Graph each inequality.

5. $y < -1$

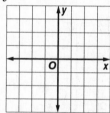

6. $y \geq x - 5$

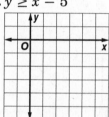

7. $y > 3x$

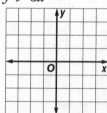

8. $y \leq 2x + 4$

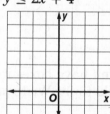

9. $y + x > 3$

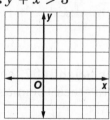

10. $y - x \geq 1$

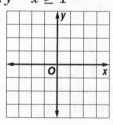

Use a graph to solve each inequality.

11. $1 > 2x + 5$

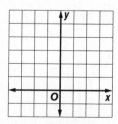

12. $7 \leq 3x + 4$

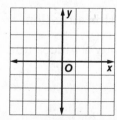

13. $-\frac{1}{2} < -\frac{1}{2}x + 1$

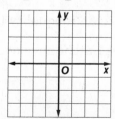

5-6 Practice

Graphing Inequalities in Two Variables

Determine which ordered pairs are part of the solution set for each inequality.

1. $3x + y \geq 6$, $\{(4, 3), (-2, 4), (-5, -3), (3, -3)\}$

2. $y \geq x + 3$, $\{(6, 3), (-3, 2), (3, -2), (4, 3)\}$

3. $3x - 2y < 5$, $\{(4, -4), (3, 5), (5, 2), (-3, 4)\}$

Graph each inequality.

4. $2y - x < -4$

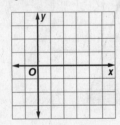

5. $2x - 2y \geq 8$

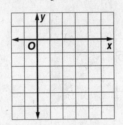

6. $3y > 2x - 3$

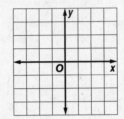

Use a graph to solve each inequality.

7. $-5 \leq x - 9$

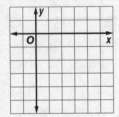

8. $6 > \frac{2}{3}x + 5$

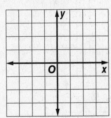

9. $\frac{1}{2} > -2x + \frac{7}{2}$

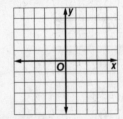

10. MOVING A moving van has an interior height of 7 feet (84 inches). You have boxes in 12 inch and 15 inch heights, and want to stack them as high as possible to fit. Write an inequality that represents this situation.

11. BUDGETING Satchi found a used bookstore that sells pre-owned videos and CDs. Videos cost $9 each, and CDs cost $7 each. Satchi can spend no more than $35.

a. Write an inequality that represents this situation.

b. Does Satchi have enough money to buy 2 videos and 3 CDs?

6-1 Skills Practice

Graphing Systems of Equations

Use the graph at the right to determine whether each system is *consistent* or *inconsistent* and if it is *independent* or *dependent*.

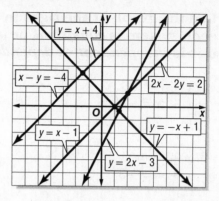

1. $y = x - 1$
   $y = -x + 1$

2. $x - y = -4$
   $y = x + 4$

3. $y = x + 4$
   $2x - 2y = 2$

4. $y = 2x - 3$
   $2x - 2y = 2$

Graph each system and determine the number of solutions that it has. If it has one solution, name it.

5. $2x - y = 1$
   $y = -3$

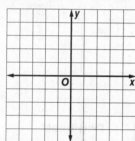

6. $x = 1$
   $2x + y = 4$

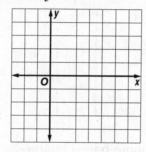

7. $3x + y = -3$
   $3x + y = 3$

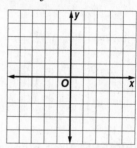

8. $y = x + 2$
   $x - y = -2$

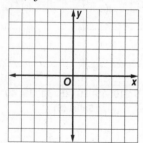

9. $x + 3y = -3$
   $x - 3y = -3$

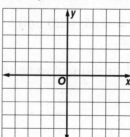

10. $y - x = -1$
    $x + y = 3$

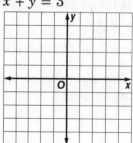

11. $x - y = 3$
    $x - 2y = 3$

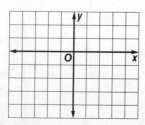

12. $x + 2y = 4$
    $y = -\frac{1}{2}x + 2$

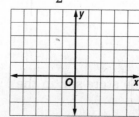

13. $y = 2x + 3$
    $3y = 6x - 6$

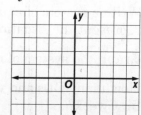

6-1 Practice

Graphing Systems of Equations

Use the graph at the right to determine whether each system is *consistent* or *inconsistent* and if it is *independent* or *dependent*.

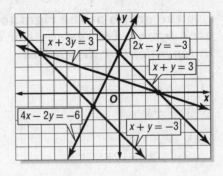

1. $x + y = 3$
$x + y = -3$

2. $2x - y = -3$
$4x - 2y = -6$

3. $x + 3y = 3$
$x + y = -3$

4. $x + 3y = 3$
$2x - y = -3$

Graph each system and determine the number of solutions that it has. If it has one solution, name it.

5. $3x - y = -2$
$3x - y = 0$

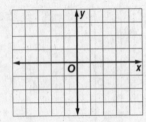

6. $y = 2x - 3$
$4x = 2y + 6$

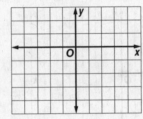

7. $x + 2y = 3$
$3x - y = -5$

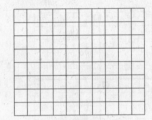

8. BUSINESS Nick plans to start a home-based business producing and selling gourmet dog treats. He figures it will cost $20 in operating costs per week plus $0.50 to produce each treat. He plans to sell each treat for $1.50.

a. Graph the system of equations $y = 0.5x + 20$ and $y = 1.5x$ to represent the situation.

b. How many treats does Nick need to sell per week to break even?

9. SALES A used book store also started selling used CDs and videos. In the first week, the store sold 40 used CDs and videos, at $4.00 per CD and $6.00 per video. The sales for both CDs and videos totaled $180.00

a. Write a system of equations to represent the situation.

b. Graph the system of equations.

c. How many CDs and videos did the store sell in the first week?

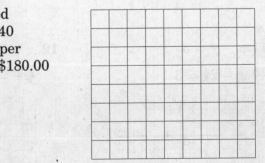

6-2 Skills Practice

Substitution

Use substitution to solve each system of equations.

1. $y = 4x$
$x + y = 5$

2. $y = 2x$
$x + 3y = -14$

3. $y = 3x$
$2x + y = 15$

4. $x = -4y$
$3x + 2y = 20$

5. $y = x - 1$
$x + y = 3$

6. $x = y - 7$
$x + 8y = 2$

7. $y = 4x - 1$
$y = 2x - 5$

8. $y = 3x + 8$
$5x + 2y = 5$

9. $2x - 3y = 21$
$y = 3 - x$

10. $y = 5x - 8$
$4x + 3y = 33$

11. $x + 2y = 13$
$3x - 5y = 6$

12. $x + 5y = 4$
$3x + 15y = -1$

13. $3x - y = 4$
$2x - 3y = -9$

14. $x + 4y = 8$
$2x - 5y = 29$

15. $x - 5y = 10$
$2x - 10y = 20$

16. $5x - 2y = 14$
$2x - y = 5$

17. $2x + 5y = 38$
$x - 3y = -3$

18. $x - 4y = 27$
$3x + y = -23$

19. $2x + 2y = 7$
$x - 2y = -1$

20. $2.5x + y = -2$
$3x + 2y = 0$

6-2 Practice

Substitution

Use substitution to solve each system of equations.

1. $y = 6x$
   $2x + 3y = -20$

2. $x = 3y$
   $3x - 5y = 12$

3. $x = 2y + 7$
   $x = y + 4$

4. $y = 2x - 2$
   $y = x + 2$

5. $y = 2x + 6$
   $2x - y = 2$

6. $3x + y = 12$
   $y = -x - 2$

7. $x + 2y = 13$
   $-2x - 3y = -18$

8. $x - 2y = 3$
   $4x - 8y = 12$

9. $x - 5y = 36$
   $2x + y = -16$

10. $2x - 3y = -24$
    $x + 6y = 18$

11. $x + 14y = 84$
    $2x - 7y = -7$

12. $0.3x - 0.2y = 0.5$
    $x - 2y = -5$

13. $0.5x + 4y = -1$
    $x + 2.5y = 3.5$

14. $3x - 2y = 11$
    $x - \frac{1}{2}y = 4$

15. $\frac{1}{2}x + 2y = 12$
    $x - 2y = 6$

16. $\frac{1}{3}x - y = 3$
    $2x + y = 25$

17. $4x - 5y = -7$
    $y = 5x$

18. $x + 3y = -4$
    $2x + 6y = 5$

19. **EMPLOYMENT** Kenisha sells athletic shoes part-time at a department store. She can earn either $500 per month plus a 4% commission on her total sales, or $400 per month plus a 5% commission on total sales.

   a. Write a system of equations to represent the situation.

   b. What is the total price of the athletic shoes Kenisha needs to sell to earn the same income from each pay scale?

   c. Which is the better offer?

20. **MOVIE TICKETS** Tickets to a movie cost $7.25 for adults and $5.50 for students. A group of friends purchased 8 tickets for $52.75.

   a. Write a system of equations to represent the situation.

   b. How many adult tickets and student tickets were purchased?

6-3 Skills Practice

Elimination Using Addition and Subtraction

Use elimination to solve each system of equations.

1. $x - y = 1$
$x + y = 3$

2. $-x + y = 1$
$x + y = 11$

3. $x + 4y = 11$
$x - 6y = 11$

4. $-x + 3y = 6$
$x + 3y = 18$

5. $3x + 4y = 19$
$3x + 6y = 33$

6. $x + 4y = -8$
$x - 4y = -8$

7. $3x + 4y = 2$
$4x - 4y = 12$

8. $3x - y = -1$
$-3x - y = 5$

9. $2x - 3y = 9$
$-5x - 3y = 30$

10. $x - y = 4$
$2x + y = -4$

11. $3x - y = 26$
$-2x - y = -24$

12. $5x - y = -6$
$-x + y = 2$

13. $6x - 2y = 32$
$4x - 2y = 18$

14. $3x + 2y = -19$
$-3x - 5y = 25$

15. $7x + 4y = 2$
$7x + 2y = 8$

16. $2x - 5y = -28$
$4x + 5y = 4$

17. The sum of two numbers is 28 and their difference is 4. What are the numbers?

18. Find the two numbers whose sum is 29 and whose difference is 15.

19. The sum of two numbers is 24 and their difference is 2. What are the numbers?

20. Find the two numbers whose sum is 54 and whose difference is 4.

21. Two times a number added to another number is 25. Three times the first number minus the other number is 20. Find the numbers.

6-3 Practice

Elimination Using Addition and Subtraction

Use elimination to solve each system of equations.

1. $x - y = 1$
$x + y = -9$

2. $p + q = -2$
$p - q = 8$

3. $4x + y = 23$
$3x - y = 12$

4. $2x + 5y = -3$
$2x + 2y = 6$

5. $3x + 2y = -1$
$4x + 2y = -6$

6. $5x + 3y = 22$
$5x - 2y = 2$

7. $5x + 2y = 7$
$-2x + 2y = -14$

8. $3x - 9y = -12$
$3x - 15y = -6$

9. $-4c - 2d = -2$
$2c - 2d = -14$

10. $2x - 6y = 6$
$2x + 3y = 24$

11. $7x + 2y = 2$
$7x - 2y = -30$

12. $4.25x - 1.28y = -9.2$
$x + 1.28y = 17.6$

13. $2x + 4y = 10$
$x - 4y = -2.5$

14. $2.5x + y = 10.7$
$2.5x + 2y = 12.9$

15. $6m - 8n = 3$
$2m - 8n = -3$

16. $4a + b = 2$
$4a + 3b = 10$

17. $-\frac{1}{3}x - \frac{4}{3}y = -2$
$\frac{1}{3}x - \frac{2}{3}y = 4$

18. $\frac{3}{4}x - \frac{1}{2}y = 8$
$\frac{3}{2}x + \frac{1}{2}y = 19$

19. The sum of two numbers is 41 and their difference is 5. What are the numbers?

20. Four times one number added to another number is 36. Three times the first number minus the other number is 20. Find the numbers.

21. One number added to three times another number is 24. Five times the first number added to three times the other number is 36. Find the numbers.

22. LANGUAGES English is spoken as the first or primary language in 78 more countries than Farsi is spoken as the first language. Together, English and Farsi are spoken as a first language in 130 countries. In how many countries is English spoken as the first language? In how many countries is Farsi spoken as the first language?

23. DISCOUNTS At a sale on winter clothing, Cody bought two pairs of gloves and four hats for $43.00. Tori bought two pairs of gloves and two hats for $30.00. What were the prices for the gloves and hats?

6-4 Skills Practice

Elimination Using Multiplication

Use elimination to solve each system of equations.

1. $x + y = -9$
   $5x - 2y = 32$

2. $3x + 2y = -9$
   $x - y = -13$

3. $2x + 5y = 3$
   $-x + 3y = -7$

4. $2x + y = 3$
   $-4x - 4y = -8$

5. $4x - 2y = -14$
   $3x - y = -8$

6. $2x + y = 0$
   $5x + 3y = 2$

7. $5x + 3y = -10$
   $3x + 5y = -6$

8. $2x + 3y = 14$
   $3x - 4y = 4$

9. $2x - 3y = 21$
   $5x - 2y = 25$

10. $3x + 2y = -26$
    $4x - 5y = -4$

11. $3x - 6y = -3$
    $2x + 4y = 30$

12. $5x + 2y = -3$
    $3x + 3y = 9$

13. Two times a number plus three times another number equals 13. The sum of the two numbers is 7. What are the numbers?

14. Four times a number minus twice another number is −16. The sum of the two numbers is −1. Find the numbers.

15. **FUNDRAISING** Trisha and Byron are washing and vacuuming cars to raise money for a class trip. Trisha raised $38 washing 5 cars and vacuuming 4 cars. Byron raised $28 by washing 4 cars and vacuuming 2 cars. Find the amount they charged to wash a car and vacuum a car.

6-4 Practice

Elimination Using Multiplication

Use elimination to solve each system of equations.

1. $2x - y = -1$
$3x - 2y = 1$

2. $5x - 2y = -10$
$3x + 6y = 66$

3. $7x + 4y = -4$
$5x + 8y = 28$

4. $2x - 4y = -22$
$3x + 3y = 30$

5. $3x + 2y = -9$
$5x - 3y = 4$

6. $4x - 2y = 32$
$-3x - 5y = -11$

7. $3x + 4y = 27$
$5x - 3y = 16$

8. $0.5x + 0.5y = -2$
$x - 0.25y = 6$

9. $2x - \frac{3}{4}y = -7$
$x + \frac{1}{2}y = 0$

10. $6x - 3y = 21$
$2x + 2y = 22$

11. $3x + 2y = 11$
$2x + 6y = -2$

12. $-3x + 2y = -15$
$2x - 4y = 26$

13. Eight times a number plus five times another number is -13. The sum of the two numbers is 1. What are the numbers?

14. Two times a number plus three times another number equals 4. Three times the first number plus four times the other number is 7. Find the numbers.

15. FINANCE Gunther invested $10,000 in two mutual funds. One of the funds rose 6% in one year, and the other rose 9% in one year. If Gunther's investment rose a total of $684 in one year, how much did he invest in each mutual fund?

16. CANOEING Laura and Brent paddled a canoe 6 miles upstream in four hours. The return trip took three hours. Find the rate at which Laura and Brent paddled the canoe in still water.

17. NUMBER THEORY The sum of the digits of a two-digit number is 11. If the digits are reversed, the new number is 45 more than the original number. Find the number.

6-5 Skills Practice

Applying Systems of Linear Equations

Determine the best method to solve each system of equations. Then solve the system.

1. $5x + 3y = 16$
$3x - 5y = -4$

2. $3x - 5y = 7$
$2x + 5y = 13$

3. $y = 3x - 24$
$5x - y = 8$

4. $-11x - 10y = 17$
$5x - 7y = 50$

5. $4x + y = 24$
$5x - y = 12$

6. $6x - y = -145$
$x = 4 - 2y$

7. VEGETABLE STAND A roadside vegetable stand sells pumpkins for $5 each and squashes for $3 each. One day they sold 6 more squash than pumpkins, and their sales totaled $98. Write and solve a system of equations to find how many pumpkins and squash they sold?

8. INCOME Ramiro earns $20 per hour during the week and $30 per hour for overtime on the weekends. One week Ramiro earned a total of $650. He worked 5 times as many hours during the week as he did on the weekend. Write and solve a system of equations to determine how many hours of overtime Ramiro worked on the weekend.

9. BASKETBALL Anya makes 14 baskets during her game. Some of these baskets were worth 2-points and others were worth 3-points. In total, she scored 30 points. Write and solve a system of equations to find how 2-points baskets she made.

6-5 Practice

Applying Systems of Linear Equations

Determine the best method to solve each system of equations. Then solve the system.

1. $1.5x - 1.9y = -29$
$x - 0.9y = 4.5$

2. $1.2x - 0.8y = -6$
$4.8x + 2.4y = 60$

3. $18x - 16y = -312$
$78x - 16y = 408$

4. $14x + 7y = 217$
$14x + 3y = 189$

5. $x = 3.6y + 0.7$
$2x + 0.2y = 38.4$

6. $5.3x - 4y = 43.5$
$x + 7y = 78$

7. BOOKS A library contains 2000 books. There are 3 times as many non-fiction books as fiction books. Write and solve a system of equations to determine the number of non-fiction and fiction books.

8. SCHOOL CLUBS The chess club has 16 members and gains a new member every month. The film club has 4 members and gains 4 new members every month. Write and solve a system of equations to find when the number of members in both clubs will be equal.

9. Tia and Ken each sold snack bars and magazine subscriptions for a school fund-raiser, as shown in the table. Tia earned \$132 and Ken earned \$190.

Item	Number Sold	
	Tia	Ken
snack bars	16	20
magazine subscriptions	4	6

a. Define variable and formulate a system of linear equation from this situation.

b. What was the price per snack bar? Determine the reasonableness of your solution.

82

6-6 Skills Practice

Organizing Data Using Matrices

State the dimensions of each matrix. Then identify the position of the circled element in each matrix.

1. $\begin{bmatrix} 0 & 3 \\ -4 & 1 \\ 2 & ⑦ \end{bmatrix}$

2. $\begin{bmatrix} 1 & -1 & 3 & ⑧ & 0 \\ 2 & 0 & -1 & 7 & -4 \\ -5 & 6 & 2 & 0 & 1 \end{bmatrix}$

3. $\begin{bmatrix} -1 & 4 \\ ⑤ & 0 \\ -2 & 7 \\ 1 & 2 \end{bmatrix}$

4. $\begin{bmatrix} 2 & -3 & 1 & 0 \\ 4 & 1 & -2 & 9 \\ 10 & 5 & 0 & ⊝1 \\ 3 & 8 & -7 & 3 \end{bmatrix}$

Perform the indicated matrix operations. If the matrix does not exist, write *impossible*.

5. $\begin{bmatrix} 5 & -1 \\ 4 & -2 \end{bmatrix} + \begin{bmatrix} 0 & 2 \\ -3 & 2 \end{bmatrix}$

6. $\begin{bmatrix} 1 & 3 \\ -4 & 9 \end{bmatrix} - \begin{bmatrix} 0 & 1 \\ 2 & 2 \end{bmatrix}$

7. $\begin{bmatrix} 9 & 1 \\ -3 & 7 \\ 0 & -2 \\ 1 & 2 \end{bmatrix} - \begin{bmatrix} 2 \\ 0 \\ 1 \\ -4 \end{bmatrix}$

8. $\begin{bmatrix} 2 & -1 & 3 \\ 4 & 0 & 1 \\ 5 & -2 & 1 \end{bmatrix} + \begin{bmatrix} 1 & 5 & -2 \\ 0 & 1 & 4 \\ -1 & 3 & 0 \end{bmatrix}$

9. $3\begin{bmatrix} 1 & -2 & 0 \\ 4 & 1 & 5 \end{bmatrix}$

10. $-2\begin{bmatrix} 8 & -3 \\ 0 & 1 \\ -2 & 5 \end{bmatrix}$

11. $5\begin{bmatrix} 1 & -4 \\ 0 & 3 \end{bmatrix}$

12. $4\begin{bmatrix} 1 & 0 & -2 & 3 & 1 \\ -4 & 5 & 2 & -1 & 0 \end{bmatrix}$

13. **WEATHER** The temperatures observed on different days in different cities are shown in the table at the right.

a. Write a matrix to organize the temperatures.

City	Monday	Tuesday	Wednesday	Thursday	Friday
Las Vegas	94°F	99°F	101°F	98°F	89°F
Phoenix	92°F	86°F	99°F	104°F	101°F

b. What are the dimensions of the matrix?

c. Which day and location had the highest temperature? lowest temperature?

6-6 Practice

Organizing Data Using Matrices

Exercises

State the dimensions of each matrix. Then identify the position of the circled element in each matrix.

1. $\begin{bmatrix} 0 & 1 & -4 & 9 \\ 2 & ⑦ & 0 & -3 \end{bmatrix}$

2. $\begin{bmatrix} 14 & -2 & 5 \\ 9 & 1 & 0 \\ -6 & ⑳ & 3 \end{bmatrix}$

3. $\begin{bmatrix} 9 & 1 & 0 & 0 \\ -3 & 2 & -4 & 1 \\ Ⓞ & 4 & 1 & 4 \end{bmatrix}$

(circled -6)

4. $[3 \quad Ⓞ & 6 \quad 1 \quad 1]$ (circled -2)

Perform the indicated matrix operations. If the matrix does not exist, write *impossible*.

5. $\begin{bmatrix} 0 & 3 \\ -1 & 8 \end{bmatrix} + \begin{bmatrix} 7 & 1 \\ 2 & 1 \end{bmatrix}$

6. $\begin{bmatrix} 3 & 2 \\ 1 & -4 \end{bmatrix} - \begin{bmatrix} 6 & -3 \\ -1 & 2 \end{bmatrix}$

7. $\begin{bmatrix} 6 & -2 & 1 \\ 3 & 4 & 0 \end{bmatrix} - \begin{bmatrix} 1 & 5 & -3 \\ 2 & -1 & 4 \end{bmatrix}$

8. $\begin{bmatrix} 7 & -1 & 3 \\ 0 & 2 & -4 \\ 3 & 1 & 5 \end{bmatrix} [3 \; -2 \; 6 \; 1 \; 1]$

9. $-2[4 \quad -1 \quad 3 \quad 0]$

10. $6\begin{bmatrix} 2 & -1 & 5 \\ -3 & 2 & -1 \end{bmatrix}$

11. $5\begin{bmatrix} 0 & 2 & -1 & 7 \\ 4 & 1 & 0 & 2 \\ -3 & 1 & 9 & -4 \\ 0 & -2 & 6 & 1 \end{bmatrix}$

12. $-2\begin{bmatrix} 2 & 11 & -5 & 4 \\ -1 & 0 & 6 & 3 \\ 9 & -2 & 1 & 0 \end{bmatrix}$

13. **FOOD SALES** The daily sales at various fast food restaurants in various cities are shown in the table below.

   a. Write a matrix to organize the sales data.

City	McPizza	Burger Hut	QuikSubs
Dulles	$25,000	$17,400	$21,000
Fitchburg	$ 3,600	$ 4,400	$ 5,900
Newton	$ 19,200	$20,100	$17,400

   b. What are the dimensions of the matrix?

   c. In which city does Burger Hut sell more food than its competitors?

6-7 Skills Practice

Using Matrices to Solve Systems of Equations

Write an augmented matrix for each system of equations.

1. $8x - y = 1$
   $x + 2y = -4$

2. $5x - 2y = 12$
   $2x + y = 8$

3. $-2x + 5y = 4$
   $4y = 8$

4. $-3x + 4y = 22$
   $2x - 3y = 6$

5. $x + 2y = 4$
   $3x - y = 5$

6. $2x - 2y = 6$
   $3x = 12$

7. $-x + 5y = 0$
   $3x + 2y = 12$

8. $x - 10y = -16$
   $3x + 2y = 6$

9. $2x = 6$
   $x + 4y = 11$

Use an augmented matrix to solve each system of equations.

10. $2x - y = -2$
   $3x + y = 17$

11. $x + 4y = 19$
   $-3x - 2y = -7$

12. $2x - y = 7$
   $-x + 3y = -11$

13. $5x - 2y = 20$
   $-x + y = -4$

14. $-2x - 4y = 2$
   $7x = 14$

15. $9x + y = 6$
   $-2x + 2y = -8$

16. $3x - 3y = 36$
   $x + 2y = 3$

17. $2x - y = 5$
   $x + y = -5$

18. $4x + y = -13$
   $2x - 5y = 21$

6-7 Practice

Using Matrices to Solve Systems of Equations

Exercises

Write an augmented matrix for each system of equations.

1. $4x - 2y = 10$
$\quad x + 8y = -22$

2. $\quad -12y = 6$
$\quad 3x + 2y = 11$

3. $\quad x + y = 10$
$\quad 2y - 3y = 0$

4. $-x + 2y = 8$
$\quad 3x - y = 5$

5. $4x - y = 11$
$\quad 2x - 3y = 3$

6. $\quad 2x = 9$
$\quad x - 5y = -5.5$

Write a system of equations for each augmented matrix.

7. $\begin{bmatrix} 2 & 0 \\ 3 & 4 \end{bmatrix} \begin{array}{|c} 8 \\ -2 \end{array}$

8. $\begin{bmatrix} 1 & 1 \\ -2 & 3 \end{bmatrix} \begin{array}{|c} 9 \\ -3 \end{array}$

9. $\begin{bmatrix} 2 & 3 \\ 1 & -4 \end{bmatrix} \begin{array}{|c} -6 \\ -14 \end{array}$

Use an augmented matrix to solve each system of equations.

10. $4x - y = 4$
$\quad 3y = 12$

11. $2x + 5y = 1$
$\quad -x - y = -2$

12. $2x + 3y = 0$
$\quad -x + 2y = 14$

13. $2x - y = 3$
$\quad 7x + y = 24$

14. $2x - y = 4$
$\quad 9x - 3y = 12$

15. $\quad 4x - y = 7$
$\quad -2x + 3y = -16$

16. COMMUTER RAIL The cost of a commuter rail ticket varies with the distance traveled. This month, Marcelo bought 5 round-trip tickets to visit his grandmother and 3 round-trip tickets to his friend's house for $31.50. Last month, Marcelo bought 2 round-trip tickets to visit his grandmother and 6 round-trip tickets to visit his friend's house for $27.00.

a. Write a system of linear equations to represent the situations.

b. Write the augmented matrix.

c. What is the cost of each type of ticket?

6-8 Skills Practice

Systems of Inequalities

Solve each system of inequalities by graphing.

1. $x > -1$
  $y \le -3$

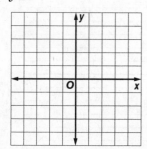

2. $y > 2$
  $x < -2$

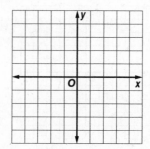

3. $y > x + 3$
  $y \le -1$

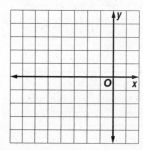

4. $x < 2$
  $y - x \le 2$

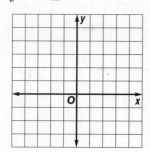

5. $x + y \le -1$
  $x + y \ge 3$

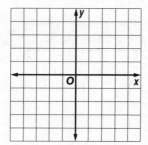

6. $y - x > 4$
  $x + y > 2$

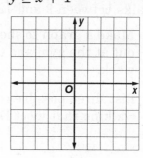

7. $y > x + 1$
  $y \ge -x + 1$

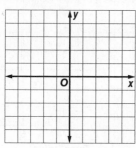

8. $y \ge -x + 2$
  $y < 2x - 2$

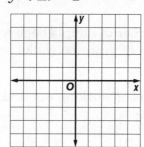

9. $y < 2x + 4$
  $y \ge x + 1$

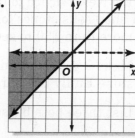

Write a system of inequalities for each graph.

10.

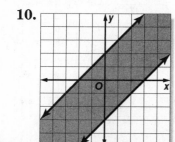

11.

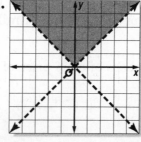

12.

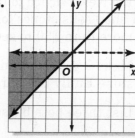

6-8 Practice

Systems of Inequalities

Solve each system of inequalities by graphing.

1. $y > x - 2$
$y \leq x$

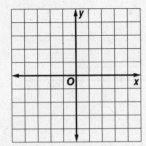

2. $y \geq x + 2$
$y > 2x + 3$

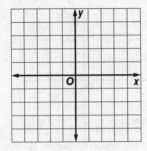

3. $x + y \geq 1$
$x + 2y > 1$

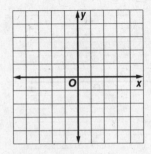

4. $y < 2x - 1$
$y > 2 - x$

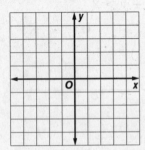

5. $y > x - 4$
$2x + y \leq 2$

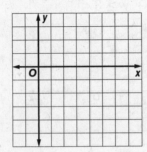

6. $2x - y \geq 2$
$x - 2y \geq 2$

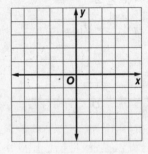

7. FITNESS Diego started an exercise program in which each week he works out at the gym between 4.5 and 6 hours and walks between 9 and 12 miles.

  a. Make a graph to show the number of hours Diego works out at the gym and the number of miles he walks per week.

  b. List three possible combinations of working out and walking that meet Diego's goals.

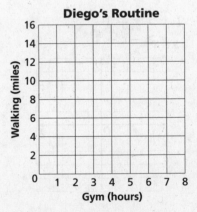

8. SOUVENIRS Emily wants to buy turquoise stones on her trip to New Mexico to give to at least 4 of her friends. The gift shop sells stones for either $4 or $6 per stone. Emily has no more than $30 to spend.

  a. Make a graph showing the numbers of each price of stone Emily can purchase.

  b. List three possible solutions.

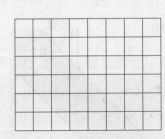

NAME _____ DATE _____ PERIOD _____

7-1 Skills Practice

Multiplying Monomials

Determine whether each expression is a monomial. Write *yes* or *no*. Explain.

1. 11

2. $a - b$

3. $\dfrac{p^2}{r^2}$

4. y

5. j^3k

6. $2a + 3b$

Simplify.

7. $a^2(a^3)(a^6)$

8. $x(x^2)(x^7)$

9. $(y^2z)(yz^2)$

10. $(\ell^2k^2)(\ell^3k)$

11. $(a^2b^4)(a^2b^2)$

12. $(cd^2)(c^3d^2)$

13. $(2x^2)(3x^5)$

14. $(5a^7)(4a^2)$

15. $(4xy^3)(3x^3y^5)$

16. $(7a^5b^2)(a^2b^3)$

17. $(-5m^3)(3m^8)$

18. $(-2c^4d)(-4cd)$

19. $(10^2)^3$

20. $(p^3)^{12}$

21. $(-6p)^2$

22. $(-3y)^3$

23. $(3pr^2)^2$

24. $(2b^3c^4)^2$

GEOMETRY Express the area of each figure as a monomial.

25.
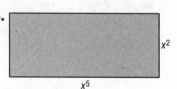
x^2 x^5

26.
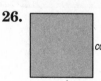
cd cd

27.
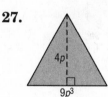
$4p$ $9p^3$

Copyright © Glencoe/McGraw-Hill, a division of The McGraw-Hill Companies, Inc.

7-1   Practice

Multiplying Monomials

Determine whether each expression is a monomial. Write *yes* or *no*. Explain your reasoning.

1. $\dfrac{21a^2}{7b}$

2. $\dfrac{b^3c^2}{2}$

Simplify each expression.

3. $(-5x^2y)(3x^4)$

4. $(2ab^2f^2)(4a^3b^2f^2)$

5. $(3ad^4)(-2a^2)$

6. $(4g^3h)(-2g^5)$

7. $(-15xy^4)\left(-\dfrac{1}{3}xy^3\right)$

8. $(-xy)^3(xz)$

9. $(-18m^2n)^2\left(-\dfrac{1}{6}mn^2\right)$

10. $(0.2a^2b^3)^2$

11. $\left(\dfrac{2}{3}p\right)^2$

12. $\left(\dfrac{1}{4}ad^3\right)^2$

13. $(0.4k^3)^3$

14. $[(4^2)^2]^2$

GEOMETRY Express the area of each figure as a monomial.

15.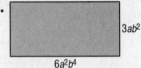
$3ab^2$
$6a^2b^4$

16.
$5x^3$

17.
$6ab^3$
$4a^2b$

GEOMETRY Express the volume of each solid as a monomial.

18.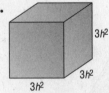
$3h^2$
$3h^2$
$3h^2$

19.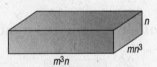
n
mn^3
m^3n

20.
$3g$
$7g^2$

21. **COUNTING** A panel of four light switches can be set in 2^4 ways. A panel of five light switches can set in twice this many ways. In how many ways can five light switches be set?

22. **HOBBIES** Tawa wants to increase her rock collection by a power of three this year and then increase it again by a power of two next year. If she has 2 rocks now, how many rocks will she have after the second year?

7-2 Skills Practice

Dividing Monomials

Simplify each expression. Assume that no denominator equals zero.

1. $\dfrac{6^5}{6^4}$

2. $\dfrac{9^{12}}{9^8}$

3. $\dfrac{x^4}{x^2}$

4. $\dfrac{r^3 t^2}{r^3 t^4}$

5. $\dfrac{m}{m^3}$

6. $\dfrac{9d^7}{3d^6}$

7. $\dfrac{12n^5}{36n}$

8. $\dfrac{w^4 x^3}{w^4 x}$

9. $\dfrac{a^3 b^5}{ab^2}$

10. $\dfrac{m^7 p^2}{m^3 p^2}$

11. $\dfrac{-21w^5 x^2}{7w^4 x^5}$

12. $\dfrac{32x^3 y^2 z^5}{-8xyz^2}$

13. $\left(\dfrac{4p^7}{7r^2}\right)^2$

14. 4^{-4}

15. 8^{-2}

16. $\left(\dfrac{5}{3}\right)^{-2}$

17. $\left(\dfrac{9}{11}\right)^{-1}$

18. $\dfrac{h^3}{h^{-6}}$

19. $k^0(k^4)(k^{-6})$

20. $k^{-1}(\ell^{-6})(m^3)$

21. $\dfrac{f^{-7}}{f^4}$

22. $\left(\dfrac{16p^5 w^2}{2p^3 w^3}\right)^0$

23. $\dfrac{f^{-5} g^4}{h^{-2}}$

24. $\dfrac{15x^6 y^{-9}}{5xy^{-11}}$

25. $\dfrac{-15t^0 u^{-1}}{5u^3}$

26. $\dfrac{48x^6 y^7 z^5}{-6xy^5 z^6}$

7-2 Practice

Dividing Monomials

Simplify each expression. Assume that no denominator equals zero.

1. $\dfrac{8^8}{8^4}$

2. $\dfrac{a^4b^6}{ab^3}$

3. $\dfrac{xy^2}{xy}$

4. $\dfrac{m^5np}{m^4p}$

5. $\dfrac{5c^2d^3}{-4c^2d}$

6. $\dfrac{8y^7z^6}{4y^6z^5}$

7. $\left(\dfrac{4f^3g}{3h^6}\right)^3$

8. $\left(\dfrac{6w^5}{7p^6r^3}\right)^2$

9. $\dfrac{-4x^2}{24x^5}$

10. $x^3(y^{-5})(x^{-8})$

11. $p(q^{-2})(r^{-3})$

12. 12^{-2}

13. $\left(\dfrac{3}{7}\right)^{-2}$

14. $\left(\dfrac{4}{3}\right)^{-4}$

15. $\dfrac{22r^3s^2}{11r^2s^{-3}}$

16. $\dfrac{-15w^0u^{-1}}{5u^3}$

17. $\dfrac{8c^3d^2f^4}{4c^{-1}d^2f^{-3}}$

18. $\left(\dfrac{x^{-3}y^5}{4^{-3}}\right)^0$

19. $\dfrac{6f^{-2}g^3h^5}{54f^{-2}g^{-5}h^3}$

20. $\dfrac{-12t^{-1}u^5x^{-4}}{2t^{-3}ux^5}$

21. $\dfrac{r^4}{(3r)^3}$

22. $\dfrac{m^{-2}n^{-5}}{(m^4n^3)^{-1}}$

23. $\dfrac{(j^{-1}k^3)^{-4}}{j^3k^3}$

24. $\dfrac{(2a^{-2}b)^{-3}}{5a^2b^4}$

25. $\left(\dfrac{q^{-1}r^3}{qr^{-2}}\right)^{-5}$

26. $\left(\dfrac{7c^{-3}d^3}{c^5dh^{-4}}\right)^{-1}$

27. $\left(\dfrac{2x^3y^2z}{3x^4yz^{-2}}\right)^{-2}$

28. **BIOLOGY** A lab technician draws a sample of blood. A cubic millimeter of the blood contains 22^3 white blood cells and 22^5 red blood cells. What is the ratio of white blood cells to red blood cells?

29. **COUNTING** The number of three-letter "words" that can be formed with the English alphabet is 26^3. The number of five-letter "words" that can be formed is 26^5. How many times more five-letter "words" can be formed than three-letter "words"?

7-3 Skills Practice

Scientific Notation

Express each number in scientific notation.

1. 3,400,000,000

2. 0.000000312

3. 2,091,000

4. 980,200,000,000,000

5. 0.00000000008

6. 0.00142

Express each number in standard form.

7. 2.1×10^5

8. 8.023×10^{-7}

9. 3.63×10^{-6}

10. 7.15×10^8

11. 1.86×10^{-4}

12. 4.9×10^5

Evaluate each product. Express the results in both scientific notation and standard form.

13. $(6.1 \times 10^5)(2 \times 10^5)$

14. $(4.4 \times 10^6)(1.6 \times 10^{-9})$

15. $(8.8 \times 10^8)(3.5 \times 10^{-13})$

16. $(1.35 \times 10^8)(7.2 \times 10^{-4})$

17. $(2.2 \times 10^{-12})(8 \times 10^6)$

18. $(3.4 \times 10^{-5})(5.4 \times 10^{-4})$

Evaluate each quotient. Express the results in both scientific notation and standard form.

19. $\dfrac{(9.2 \times 10^{-8})}{(2 \times 10^{-6})}$

20. $\dfrac{(4.8 \times 10^4)}{(3 \times 10^{-5})}$

21. $\dfrac{(1.161 \times 10^{-9})}{(4.3 \times 10^{-6})}$

22. $\dfrac{(4.625 \times 10^{10})}{(1.25 \times 10^4)}$

23. $\dfrac{(2.376 \times 10^{-4})}{(7.2 \times 10^{-8})}$

24. $\dfrac{(8.74 \times 10^{-3})}{(1.9 \times 10^5)}$

7-3 Practice

Scientific Notation

Express each number in scientific notation.

1. 1,900,000

2. 0.000704

3. 50,040,000,000

4. 0.0000000661

Express each number in standard form.

5. 5.3×10^7

6. 1.09×10^{-4}

7. 9.13×10^3

8. 7.902×10^{-6}

Evaluate each product. Express the results in both scientific notation and standard form.

9. $(4.8 \times 10^4)(6 \times 10^6)$

10. $(7.5 \times 10^{-5})(3.2 \times 10^7)$

11. $(2.06 \times 10^4)(5.5 \times 10^{-9})$

12. $(8.1 \times 10^{-6})(1.96 \times 10^{11})$

13. $(5.29 \times 10^8)(9.7 \times 10^4)$

14. $(1.45 \times 10^{-6})(7.2 \times 10^{-5})$

Evaluate each quotient. Express the results in both scientific notation and standard form.

15. $\dfrac{(4.2 \times 10^5)}{(3 \times 10^{-3})}$

16. $\dfrac{(1.76 \times 10^{-11})}{(2.2 \times 10^{-5})}$

17. $\dfrac{(7.05 \times 10^{12})}{(9.4 \times 10^7)}$

18. $\dfrac{(2.04 \times 10^{-4})}{(3.4 \times 10^5)}$

19. GRAVITATION Issac Newton's theory of universal gravitation states that the equation $F = G\dfrac{m_1 m_2}{r^2}$ can be used to calculate the amount of gravitational force in newtons between two point masses m_1 and m_2 separated by a distance r. G is a constant equal to 6.67×10^{-11} N m^2 kg^{-2}. The mass of the earth m_1 is equal to 5.97×10^{24} kg, the mass of the moon m_2 is equal to 7.36×10^{22} kg, and the distance r between the two is 384,000,000 m.

a. Express the distance r in scientific notation.

b. Compute the amount of gravitational force between the earth and the moon. Express your answer in scientific notation.

7-4 Skills Practice

Polynomials

Determine whether each expression is a polynomial. If so, identify the polynomial as a *monomial*, *binomial*, or *trinomial*.

1. $5mt + t^2$

2. $4by + 2b - by$

3. -32

4. $\dfrac{3x}{7}$

5. $5x^2 - 3x^{-4}$

6. $2c^2 + 8c + 9 - 3$

Find the degree of each polynomial.

7. 12

8. $3r^4$

9. $b + 6$

10. $4a^3 - 2a$

11. $5abc - 2b^2 + 1$

12. $8x^5y^4 - 2x^8$

Write each polynomial in standard form. Identify the leading coefficient.

13. $3x + 1 + 2x^2$

14. $5x - 6 + 3x^2$

15. $9x^2 + 2 + x^3 + x$

16. $-3 + 3x^3 - x^2 + 4x$

17. $x^2 + 3x^3 + 27 - x$

18. $25 - x^3 + x$

19. $x - 3x^2 + 4 + 5x^3$

20. $x^2 + 64 - x + 7x^3$

21. $6x^3 - 7x^5 + x - 2x^2 + 1$

22. $4 - x + 3x^3 - 2x^2$

23. $13 - 4x^9 + x^3$

24. $17x^5 - 5x^{17} + 2$

7-4 Practice

Polynomials

Determine whether each expression is a polynomial. If so, identify the polynomial as a *monomial*, *binomial*, or *trinomial*.

1. $7a^2b + 3b^2 - a^2b$

2. $\frac{1}{5}y^3 + y^2 - 9$

3. $6g^2h^3k$

Find the degree of each polynomial.

4. $x + 3x^4 - 21x^2 + x^3$

5. $3g^2h^3 + g^3h$

6. $-2x^2y + 3xy^3 + x^2$

7. $5n^3m - 2m^3 + n^2m^4 + n^2$

8. $a^3b^2c + 2a^5c + b^3c^2$

9. $10r^2t^2 + 4rt^2 - 5r^3t^2$

Write each polynomial in standard form. Identify the leading coefficient.

10. $8x^2 - 15 + 5x^5$

11. $10x - 7 + x^4 + 4x^3$

12. $13x^2 - 5 + 6x^3 - x$

13. $4x + 2x^5 - 6x^3 + 2$

GEOMETRY Write a polynomial to represent the area of each shaded region.

14.

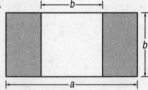

15.

16. MONEY Write a polynomial to represent the value of t ten-dollar bills, f fifty-dollar bills, and h one-hundred-dollar bills.

17. GRAVITY The height above the ground of a ball thrown up with a velocity of 96 feet per second from a height of 6 feet is $6 + 96t - 16t^2$ feet, where t is the time in seconds. According to this model, how high is the ball after 7 seconds? Explain.

7-5 Skills Practice

Adding and Subtracting Polynomials

Find each sum or difference.

1. $(2x + 3y) + (4x + 9y)$

2. $(6s + 5t) + (4t + 8s)$

3. $(5a + 9b) - (2a + 4b)$

4. $(11m - 7n) - (2m + 6n)$

5. $(m^2 - m) + (2m + m^2)$

6. $(x^2 - 3x) - (2x^2 + 5x)$

7. $(d^2 - d + 5) - (2d + 5)$

8. $(2h^2 - 5h) + (7h - 3h^2)$

9. $(5f + g - 2) + (-2f + 3)$

10. $(6k^2 + 2k + 9) + (4k^2 - 5k)$

11. $(x^3 - x + 1) - (3x - 1)$

12. $(b^2 + ab - 2) - (2b^2 + 2ab)$

13. $(7z^2 + 4 - z) - (-5 + 3z^2)$

14. $(5 + 4n + 2t) + (-6t - 8)$

15. $(4t^2 + 2) + (-4 + 2t)$

16. $(3g^3 + 7g) - (4g + 8g^3)$

17. $(2a^2 + 8a + 4) - (a^2 - 3)$

18. $(3x^2 - 7x + 5) - (-x^2 + 4x)$

19. $(7y^2 + y + 1) - (-4y + 3y^2 - 3)$

20. $(2c^2 + 7c + 4) + (c^2 + 1 - 9c)$

21. $(n^2 + 3n + 2) - (2n^2 - 6n - 2)$

22. $(a^2 + ab - 3b^2) + (b^2 + 4a^2 - ab)$

23. $(\ell^2 - 5\ell - 6) + (2\ell^2 + 5 + \ell)$

24. $(2m^2 + 5m + 1) - (4m^2 - 3m - 3)$

25. $(x^2 - 6x + 2) - (-5x^2 + 7x - 4)$

26. $(5b^2 - 9b - 5) + (b^2 - 6 + 2b)$

27. $(2x^2 - 6x - 2) + (x^2 + 4x) + (3x^2 + x + 5)$

7-5 Practice

Adding and Subtracting Polynomials

Find each sum or difference.

1. $(4y + 5) + (-7y - 1)$

2. $(-x^2 + 3x) - (5x + 2x^2)$

3. $(4k^2 + 8k + 2) - (2k + 3)$

4. $(2m^2 + 6m) + (m^2 - 5m + 7)$

5. $(5a^2 + 6a + 2) - (7a^2 - 7a + 5)$

6. $(-4p^2 - p + 9) + (p^2 + 3p - 1)$

7. $(x^3 - 3x + 1) - (x^3 + 7 - 12x)$

8. $(6x^2 - x + 1) - (-4 + 2x^2 + 8x)$

9. $(4y^2 + 2y - 8) - (7y^2 + 4 - y)$

10. $(w^2 - 4w - 1) + (-5 + 5w^2 - 3w)$

11. $(4u^2 - 2u - 3) + (3u^2 - u + 4)$

12. $(5b^2 - 8 + 2b) - (b + 9b^2 + 5)$

13. $(4d^2 + 2d + 2) + (5d^2 - 2 - d)$

14. $(8x^2 + x - 6) - (-x^2 + 2x - 3)$

15. $(3h^2 + 7h - 1) - (4h + 8h^2 + 1)$

16. $(4m^2 - 3m + 10) + (m^2 + m - 2)$

17. $(x^2 + y^2 - 6) - (5x^2 - y^2 - 5)$

18. $(7t^2 + 2 - t) + (t^2 - 7 - 2t)$

19. $(k^3 - 2k^2 + 4k + 6) - (-4k + k^2 - 3)$

20. $(9j^2 + j + jk) + (-3j^2 - jk - 4j)$

21. $(2x + 6y - 3z) + (4x + 6z - 8y) + (x - 3y + z)$

22. $(6f^2 - 7f - 3) - (5f^2 - 1 + 2f) - (2f^2 - 3 + f)$

23. BUSINESS The polynomial $s^3 - 70s^2 + 1500s - 10{,}800$ models the profit a company makes on selling an item at a price s. A second item sold at the same price brings in a profit of $s^3 - 30s^2 + 450s - 5000$. Write a polynomial that expresses the total profit from the sale of both items.

24. GEOMETRY The measures of two sides of a triangle are given. If P is the perimeter, and $P = 10x + 5y$, find the measure of the third side.

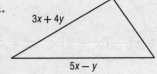

$3x + 4y$

$5x - y$

7-6 Skills Practice

Multiplying a Polynomial by a Monomial

Find each product.

1. $a(4a + 3)$

2. $-c(11c + 4)$

3. $x(2x - 5)$

4. $2y(y - 4)$

5. $-3n(n^2 + 2n)$

6. $4h(3h - 5)$

7. $3x(5x^2 - x + 4)$

8. $7c(5 - 2c^2 + c^3)$

9. $-4b(1 - 9b - 2b^2)$

10. $6y(-5 - y + 4y^2)$

11. $2m^2(2m^2 + 3m - 5)$

12. $-3n^2(-2n^2 + 3n + 4)$

Simplify each expression.

13. $w(3w + 2) + 5w$

14. $f(5f - 3) - 2f$

15. $-p(2p - 8) - 5p$

16. $y^2(-4y + 5) - 6y^2$

17. $2x(3x^2 + 4) - 3x^3$

18. $4a(5a^2 - 4) + 9a$

19. $4b(-5b - 3) - 2(b^2 - 7b - 4)$

20. $3m(3m + 6) - 3(m^2 + 4m + 1)$

Solve each equation.

21. $3(a + 2) + 5 = 2a + 4$

22. $2(4x + 2) - 8 = 4(x + 3)$

23. $5(y + 1) + 2 = 4(y + 2) - 6$

24. $4(b + 6) = 2(b + 5) + 2$

25. $6(m - 2) + 14 = 3(m + 2) - 10$

26. $3(c + 5) - 2 = 2(c + 6) + 2$

7-6 Practice

Multiplying a Polynomial by a Monomial

Find each product.

1. $2h(-7h^2 - 4h)$

2. $6pq(3p^2 + 4q)$

3. $5jk(3jk + 2k)$

4. $-3rt(-2t^2 + 3r)$

5. $-\dfrac{1}{4}m(8m^2 + m - 7)$

6. $-\dfrac{2}{3}n^2(-9n^2 + 3n + 6)$

Simplify each expression.

7. $-2\ell(3\ell - 4) + 7\ell$

8. $5w(-7w + 3) + 2w(-2w^2 + 19w + 2)$

9. $6t(2t - 3) - 5(2t^2 + 9t - 3)$

10. $-2(3m^3 + 5m + 6) + 3m(2m^2 + 3m + 1)$

11. $-3g(7g - 2) + 3(g^2 + 2g + 1) - 3g(-5g + 3)$

Solve each equation.

12. $5(2t - 1) + 3 = 3(3t + 2)$

13. $3(3u + 2) + 5 = 2(2u - 2)$

14. $4(8n + 3) - 5 = 2(6n + 8) + 1$

15. $8(3b + 1) = 4(b + 3) - 9$

16. $t(t + 4) - 1 = t(t + 2) + 2$

17. $u(u - 5) + 8u = u(u + 2) - 4$

18. NUMBER THEORY Let x be an integer. What is the product of twice the integer added to three times the next consecutive integer?

19. INVESTMENTS Kent invested $5000 in a retirement plan. He allocated x dollars of the money to a bond account that earns 4% interest per year and the rest to a traditional account that earns 5% interest per year.

  a. Write an expression that represents the amount of money invested in the traditional account.

  b. Write a polynomial model in simplest form for the total amount of money T Kent has invested after one year. (*Hint:* Each account has $A + IA$ dollars, where A is the original amount in the account and I is its interest rate.)

  c. If Kent put $500 in the bond account, how much money does he have in his retirement plan after one year?

7-7 Skills Practice

Multiplying Polynomials

Find each product.

1. $(m + 4)(m + 1)$

2. $(x + 2)(x + 2)$

3. $(b + 3)(b + 4)$

4. $(t + 4)(t - 3)$

5. $(r + 1)(r - 2)$

6. $(n - 5)(n + 1)$

7. $(3c + 1)(c - 2)$

8. $(2x - 6)(x + 3)$

9. $(d - 1)(5d - 4)$

10. $(2\ell + 5)(\ell - 4)$

11. $(3n - 7)(n + 3)$

12. $(q + 5)(5q - 1)$

13. $(3b + 3)(3b - 2)$

14. $(2m + 2)(3m - 3)$

15. $(4c + 1)(2c + 1)$

16. $(5a - 2)(2a - 3)$

17. $(4h - 2)(4h - 1)$

18. $(x - y)(2x - y)$

19. $(w + 4)(w^2 + 3w - 6)$

20. $(t + 1)(t^2 + 2t + 4)$

21. $(k + 4)(k^2 + 3k - 6)$

22. $(m + 3)(m^2 + 3m + 5)$

7-7 Practice

Multiplying Polynomials

Find each product.

1. $(q + 6)(q + 5)$

2. $(x + 7)(x + 4)$

3. $(n - 4)(n - 6)$

4. $(a + 5)(a - 6)$

5. $(4b + 6)(b - 4)$

6. $(2x - 9)(2x + 4)$

7. $(6a - 3)(7a - 4)$

8. $(2x - 2)(5x - 4)$

9. $(3a - b)(2a - b)$

10. $(4g + 3h)(2g + 3h)$

11. $(m + 5)(m^2 + 4m - 8)$

12. $(t + 3)(t^2 + 4t + 7)$

13. $(2h + 3)(2h^2 + 3h + 4)$

14. $(3d + 3)(2d^2 + 5d - 2)$

15. $(3q + 2)(9q^2 - 12q + 4)$

16. $(3r + 2)(9r^2 + 6r + 4)$

17. $(3n^2 + 2n - 1)(2n^2 + n + 9)$

18. $(2t^2 + t + 3)(4t^2 + 2t - 2)$

19. $(2x^2 - 2x - 3)(2x^2 - 4x + 3)$

20. $(3y^2 + 2y + 2)(3y^2 - 4y - 5)$

GEOMETRY Write an expression to represent the area of each figure.

21.

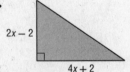

22.

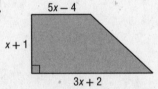

23. NUMBER THEORY Let x be an even integer. What is the product of the next two consecutive even integers?

24. GEOMETRY The volume of a rectangular pyramid is one third the product of the area of its base and its height. Find an expression for the volume of a rectangular pyramid whose base has an area of $3x^2 + 12x + 9$ square feet and whose height is $x + 3$ feet.

7-8 Skills Practice

Special Products

Find each product.

1. $(n + 3)^2$

2. $(x + 4)(x + 4)$

3. $(y - 7)^2$

4. $(t - 3)(t - 3)$

5. $(b + 1)(b - 1)$

6. $(a - 5)(a + 5)$

7. $(p - 4)^2$

8. $(z + 3)(z - 3)$

9. $(\ell + 2)(\ell + 2)$

10. $(r - 1)(r - 1)$

11. $(3g + 2)(3g - 2)$

12. $(2m - 3)(2m + 3)$

13. $(6 + u)^2$

14. $(r + t)^2$

15. $(3q + 1)(3q - 1)$

16. $(c - d)^2$

17. $(2k - 2)^2$

18. $(w + 3h)^2$

19. $(3p - 4)(3p + 4)$

20. $(t + 2u)^2$

21. $(x - 4y)^2$

22. $(3b + 7)(3b - 7)$

23. $(3y - 3g)(3y + 3g)$

24. $(n^2 + r^2)^2$

25. $(2k + m^2)^2$

26. $(3t^2 - n)^2$

27. GEOMETRY The length of a rectangle is the sum of two whole numbers. The width of the rectangle is the difference of the same two whole numbers. Using these facts, write a verbal expression for the area of the rectangle.

7-8 Practice

Special Products

Find each product.

1. $(n + 9)^2$

2. $(q + 8)^2$

3. $(x - 10)^2$

4. $(r - 11)^2$

5. $(p + 7)^2$

6. $(b + 6)(b - 6)$

7. $(z + 13)(z - 13)$

8. $(4j + 2)^2$

9. $(5w - 4)^2$

10. $(6h - 1)^2$

11. $(3m + 4)^2$

12. $(7v - 2)^2$

13. $(7k + 3)(7k - 3)$

14. $(4d - 7)(4d + 7)$

15. $(3g + 9h)(3g - 9h)$

16. $(4q + 5t)(4q - 5t)$

17. $(a + 6u)^2$

18. $(5r + s)^2$

19. $(6h - m)^2$

20. $(k - 6y)^2$

21. $(u - 7p)^2$

22. $(4b - 7v)^2$

23. $(6n + 4p)^2$

24. $(5q + 6t)^2$

25. $(6a - 7b)(6a + 7b)$

26. $(8h + 3d)(8h - 3d)$

27. $(9x + 2y^2)^2$

28. $(3p^3 + 2m)^2$

29. $(5a^2 - 2b)^2$

30. $(4m^3 - 2t)^2$

31. $(6b^3 - g)^2$

32. $(2b^2 - g)(2b^2 + g)$

33. $(2v^2 + 3x^2)(2v^2 + 3x^2)$

34. **GEOMETRY** Janelle wants to enlarge a square graph that she has made so that a side of the new graph will be 1 inch more than twice the original side g. What trinomial represents the area of the enlarged graph?

35. **GENETICS** In a guinea pig, pure black hair coloring B is dominant over pure white coloring b. Suppose two hybrid Bb guinea pigs, with black hair coloring, are bred.

   a. Find an expression for the genetic make-up of the guinea pig offspring.

   b. What is the probability that two hybrid guinea pigs with black hair coloring will produce a guinea pig with white hair coloring?

8-1 Skills Practice

Monomials and Factoring

Factor each monomial completely.

1. $10a^4$

2. $-27x^3y^2$

3. $28pr^2$

4. $44m^2np^3$

5. $9x^3y^2$

6. $-17ab^2f$

7. $42g^2$

8. $36tu^2$

9. $-4a$

10. $-10x^4yz^2$

Find the GCF of each set of monomials.

11. $16f$, $21ab^2$

12. $18t$, $48t^4$

13. $32xyz$, $48xy^4$

14. $12m^3p^2$, $44mp^3$

15. $4q^2r^2t^2$, $9q^3r^3t^3$

16. $14ab^5$, $7a^2b^3c$

17. $51xyz^2$, $68x^2yz^2$

18. $12t^7u^3$, $18t^3u^7$

19. $11a^4b^3$, $44a^2b^5$

20. $18r^3t$, $26qr^2t^4$

8-1 Practice

Monomials and Factoring

Factor each monomial completely.

1. $30d^5$

2. $-72mp$

3. $81b^2c^3$

4. $145abc^3$

5. $168nq^2r$

6. $-121x^2yz^2$

7. $-14f^2g^2$

8. $-77w^4$

Find the GCF of each set of monomials.

9. $24fg^5$, $56f^3g$

10. $72r^2t^2$, $36rt^3$

11. $15a^2b$, $35ab^2$

12. $28k^3n^2$, $45pr^2$

13. $40xy^2$, $56x^3y^2$, $124x^2y^3$

14. $88a^3d$, $40a^2d^2$, $32a^2d$

15. **GEOMETRY** The area of a rectangle is 84 square inches. Its length and width are both whole numbers.

   a. What is the minimum perimeter of the rectangle?

   b. What is the maximum perimeter of the rectangle?

16. **RENOVATION** Ms. Baxter wants to tile a wall to serve as a splashguard above a basin in the basement. She plans to use equal-sized tiles to cover an area that measures 48 inches by 36 inches.

   a. What is the maximum-size square tile Ms. Baxter can use and not have to cut any of the tiles?

   b. How many tiles of this size will she need?

8-2 Skills Practice

Using the Distributive Property

Factor each polynomial.

1. $7x + 49$

2. $8m - 6$

3. $5a^2 - 15$

4. $10q - 25q^2$

5. $8ax - 56a$

6. $81r + 48rt$

7. $t^2h + 3t$

8. $a^2b^2 + a$

9. $x + x^2y + x^3y^2$

10. $3p^2r^2 + 6pr + p$

11. $4a^2b^2 + 16ab + 12a$

12. $10h^3n^3 - 2hn^2 + 14hn$

13. $x^2 + 3x + x + 3$

14. $b^2 - 2b + 3b - 6$

15. $2j^2 + 2j + 3j + 3$

16. $2a^2 - 4a + a - 2$

17. $6t^2 - 4t - 3t + 2$

18. $9x^2 - 3xy + 6x - 2y$

Solve each equation. Check your solutions.

19. $x(x - 8) = 0$

20. $b(b + 12) = 0$

21. $(m - 3)(m + 5) = 0$

22. $(a - 9)(2a + 1) = 0$

23. $x^2 - 5x = 0$

24. $y^2 + 3y = 0$

25. $3a^2 = 6a$

26. $2x^2 = 3x$

8-2 Practice

Using the Distributive Property

Factor each polynomial.

1. $64 - 40ab$

2. $4d^2 + 16$

3. $6r^2t - 3rt^2$

4. $15ad + 30a^2d^2$

5. $32a^2 + 24b^2$

6. $36xy^2 - 48x^2y$

7. $30x^3y + 35x^2y^2$

8. $9a^3d^2 - 6ad^3$

9. $75b^2g^3 + 60bg^3$

10. $8p^2r^2 - 24pr^3 + 16pr$

11. $5x^3y^2 + 10x^2y + 25x$

12. $9ax^3 + 18bx^2 + 24cx$

13. $x^2 + 4x + 2x + 8$

14. $2a^2 + 3a + 6a + 9$

15. $4b^2 - 12b + 2b - 6$

16. $6xy - 8x + 15y - 20$

17. $-6mp + 4m + 18p - 12$

18. $12a^2 - 15ab - 16a + 20b$

Solve each equation. Check your solutions.

19. $x(x - 32) = 0$

20. $4b(b + 4) = 0$

21. $(y - 3)(y + 2) = 0$

22. $(a + 6)(3a - 7) = 0$

23. $(2y + 5)(y - 4) = 0$

24. $(4y + 8)(3y - 4) = 0$

25. $2z^2 + 20z = 0$

26. $8p^2 - 4p = 0$

27. $9x^2 = 27x$

28. $18x^2 = 15x$

29. $14x^2 = -21x$

30. $8x^2 = -26x$

31. LANDSCAPING A landscaping company has been commissioned to design a triangular flower bed for a mall entrance. The final dimensions of the flower bed have not been determined, but the company knows that the height will be two feet less than the base. The area of the flower bed can be represented by the equation $A = \frac{1}{2}b^2 - b$.

   a. Write this equation in factored form.

   b. Suppose the base of the flower bed is 16 feet. What will be its area?

32. PHYSICAL SCIENCE Mr. Alim's science class launched a toy rocket from ground level with an initial upward velocity of 60 feet per second. The height h of the rocket in feet above the ground after t seconds is modeled by the equation $h = 60t - 16t^2$. How long was the rocket in the air before it returned to the ground?

8-3 Skills Practice

Quadratic Equations: $x^2 + bx + c = 0$

Factor each polynomial.

1. $t^2 + 8t + 12$

2. $n^2 + 7n + 12$

3. $p^2 + 9p + 20$

4. $h^2 + 9h + 18$

5. $n^2 + 3n - 18$

6. $x^2 + 2x - 8$

7. $y^2 - 5y - 6$

8. $g^2 + 3g - 10$

9. $r^2 + 4r - 12$

10. $x^2 - x - 12$

11. $w^2 - w - 6$

12. $y^2 - 6y + 8$

13. $x^2 - 8x + 15$

14. $b^2 - 9b + 8$

15. $t^2 - 15t + 56$

16. $-4 - 3m + m^2$

Solve each equation. Check the solutions.

17. $x^2 - 6x + 8 = 0$

18. $b^2 - 7b + 12 = 0$

19. $m^2 + 5m + 6 = 0$

20. $d^2 + 7d + 10 = 0$

21. $y^2 - 2y - 24 = 0$

22. $p^2 - 3p = 18$

23. $h^2 + 2h = 35$

24. $a^2 + 14a = -45$

25. $n^2 - 36 = 5n$

26. $w^2 + 30 = 11w$

8-3 Practice

Quadratic Equations: $x^2 + bx + c = 0$

Factor each polynomial.

1. $a^2 + 10a + 24$

2. $h^2 + 12h + 27$

3. $x^2 + 14x + 33$

4. $g^2 - 2g - 63$

5. $w^2 + w - 56$

6. $y^2 + 4y - 60$

7. $b^2 + 4b - 32$

8. $n^2 - 3n - 28$

9. $t^2 + 4t - 45$

10. $z^2 - 11z + 30$

11. $d^2 - 16d + 63$

12. $x^2 - 11x + 24$

13. $q^2 - q - 56$

14. $x^2 - 6x - 55$

15. $32 + 18r + r^2$

16. $48 - 16g + g^2$

17. $j^2 - 9jk - 10k^2$

18. $m^2 - mv - 56v^2$

Solve each equation. Check the solutions.

19. $x^2 + 17x + 42 = 0$

20. $p^2 + 5p - 84 = 0$

21. $k^2 + 3k - 54 = 0$

22. $b^2 - 12b - 64 = 0$

23. $n^2 + 4n = 32$

24. $h^2 - 17h = -60$

25. $t^2 - 26t = 56$

26. $z^2 - 14z = 72$

27. $y^2 - 84 = 5y$

28. $80 + a^2 = 18a$

29. $u^2 = 16u + 36$

30. $17r + r^2 = -52$

31. Find all values of k so that the trinomial $x^2 + kx - 35$ can be factored using integers.

32. CONSTRUCTION A construction company is planning to pour concrete for a driveway. The length of the driveway is 16 feet longer than its width w.

 a. Write an expression for the area of the driveway.

 b. Find the dimensions of the driveway if it has an area of 260 square feet.

33. WEB DESIGN Janeel has a 10-inch by 12-inch photograph. She wants to scan the photograph, then reduce the result by the same amount in each dimension to post on her Web site. Janeel wants the area of the image to be one eighth that of the original photograph.

 a. Write an equation to represent the area of the reduced image.

 b. Find the dimensions of the reduced image.

8-4 Skills Practice

Quadratic Equations: $ax^2 + bx + c = 0$

Factor each polynomial, if possible. If the polynomial cannot be factored using integers, write *prime*.

1. $2x^2 + 5x + 2$

2. $3n^2 + 5n + 2$

3. $2t^2 + 9t - 5$

4. $3g^2 - 7g + 2$

5. $2t^2 - 11t + 15$

6. $2x^2 + 3x - 6$

7. $2y^2 + y - 1$

8. $4h^2 + 8h - 5$

9. $4x^2 - 3x - 3$

10. $4b^2 + 15b - 4$

11. $9p^2 + 6p - 8$

12. $6q^2 - 13q + 6$

13. $3a^2 + 30a + 63$

14. $10w^2 - 19w - 15$

Solve each equation. Check the solutions.

15. $2x^2 + 7x + 3 = 0$

16. $3w^2 + 14w + 8 = 0$

17. $3n^2 - 7n + 2 = 0$

18. $5d^2 - 22d + 8 = 0$

19. $6h^2 + 8h + 2 = 0$

20. $8p^2 - 16p = 10$

21. $9y^2 + 18y - 12 = 6y$

22. $4a^2 - 16a = -15$

23. $10b^2 - 15b = 8b - 12$

24. $6d^2 + 21d = 10d + 35$

8-4 Practice

Quadratic Equations: $ax^2 + bx + c = 0$

Factor each polynomial, if possible. If the polynomial cannot be factored using integers, write *prime*.

1. $2b^2 + 10b + 12$

2. $3g^2 + 8g + 4$

3. $4x^2 + 4x - 3$

4. $8b^2 - 5b - 10$

5. $6m^2 + 7m - 3$

6. $10d^2 + 17d - 20$

7. $6a^2 - 17a + 12$

8. $8w^2 - 18w + 9$

9. $10x^2 - 9x + 6$

10. $15n^2 - n - 28$

11. $10x^2 + 21x - 10$

12. $9r^2 + 15r + 6$

13. $12y^2 - 4y - 5$

14. $14k^2 - 9k - 18$

15. $8z^2 + 20z - 48$

16. $12q^2 + 34q - 28$

17. $18h^2 + 15h - 18$

18. $12p^2 - 22p - 20$

Solve each equation. Check the solutions.

19. $3h^2 + 2h - 16 = 0$

20. $15n^2 - n = 2$

21. $8q^2 - 10q + 3 = 0$

22. $6b^2 - 5b = 4$

23. $10r^2 - 21r = -4r + 6$

24. $10g^2 + 10 = 29g$

25. $6y^2 = -7y - 2$

26. $9z^2 = -6z + 15$

27. $12k^2 + 15k = 16k + 20$

28. $12x^2 - 1 = -x$

29. $8a^2 - 16a = 6a - 12$

30. $18a^2 + 10a = -11a + 4$

31. DIVING Lauren dove into a swimming pool from a 15-foot-high diving board with an initial upward velocity of 8 feet per second. Find the time t in seconds it took Lauren to enter the water. Use the model for vertical motion given by the equation $h = -16t^2 + vt + s$, where h is height in feet, t is time in seconds, v is the initial upward velocity in feet per second, and s is the initial height in feet. (*Hint:* Let $h = 0$ represent the surface of the pool.)

32. BASEBALL Brad tossed a baseball in the air from a height of 6 feet with an initial upward velocity of 14 feet per second. Enrique caught the ball on its way down at a point 4 feet above the ground. How long was the ball in the air before Enrique caught it? Use the model of vertical motion from Exercise 31.

8-5 Skills Practice

Quadratic Equations: Differences of Squares

Factor each polynomial, if possible. If the polynomial cannot be factored, write *prime*.

1. $a^2 - 4$

2. $n^2 - 64$

3. $1 - 49d^2$

4. $-16 + p^2$

5. $k^2 + 25$

6. $36 - 100w^2$

7. $t^2 - 81u^2$

8. $4h^2 - 25g^2$

9. $64m^2 - 9y^2$

10. $4c^2 - 5d^2$

11. $-49r^2 + 4t^2$

12. $8x^2 - 72p^2$

13. $20q^2 - 5r^2$

14. $32a^2 - 50b^2$

Solve each equation by factoring. Check the solutions.

15. $16x^2 - 9 = 0$

16. $25p^2 - 16 = 0$

17. $36q^2 - 49 = 0$

18. $81 - 4b^2 = 0$

19. $16d^2 = 4$

20. $18a^2 = 8$

21. $n^2 - \dfrac{9}{25} = 0$

22. $k^2 - \dfrac{49}{64} = 0$

23. $\dfrac{1}{25}h^2 - 16 = 0$

24. $\dfrac{1}{16}y^2 = 81$

8-5 Practice

Quadratic Equations: Differences of Squares

Factor each polynomial, if possible. If the polynomial cannot be factored, write prime.

1. $k^2 - 100$

2. $81 - r^2$

3. $16p^2 - 36$

4. $4x^2 + 25$

5. $144 - 9f^2$

6. $36g^2 - 49h^2$

7. $121m^2 - 144p^2$

8. $32 - 8y^2$

9. $24a^2 - 54b^2$

10. $32t^2 - 18u^2$

11. $9d^2 - 32$

12. $36z^3 - 9z$

13. $45q^3 - 20q$

14. $100b^3 - 36b$

15. $3t^4 - 48t^2$

Solve each equation by factoring. Check your solutions.

16. $4y^2 = 81$

17. $64p^2 = 9$

18. $98b^2 - 50 = 0$

19. $32 - 162k^2 = 0$

20. $t^2 - \dfrac{64}{121} = 0$

21. $\dfrac{16}{49} - v^2 = 0$

22. $\dfrac{1}{36}x^2 - 25 = 0$

23. $27h^3 = 48h$

24. $75g^3 = 147g$

25. EROSION A rock breaks loose from a cliff and plunges toward the ground 400 feet below. The distance d that the rock falls in t seconds is given by the equation $d = 16t^2$. How long does it take the rock to hit the ground?

26. FORENSICS Mr. Cooper contested a speeding ticket given to him after he applied his brakes and skidded to a halt to avoid hitting another car. In traffic court, he argued that the length of the skid marks on the pavement, 150 feet, proved that he was driving under the posted speed limit of 65 miles per hour. The ticket cited his speed at 70 miles per hour. Use the formula $\dfrac{1}{24}s^2 = d$, where s is the speed of the car and d is the length of the skid marks, to determine Mr. Cooper's speed when he applied the brakes. Was Mr. Cooper correct in claiming that he was not speeding when he applied the brakes?

8-6 Skills Practice

Quadratic Equations: Perfect Squares

Determine whether each trinomial is a perfect square trinomial. Write *yes* or *no*. If so, factor it.

1. $m^2 - 6m + 9$

2. $r^2 + 4r + 4$

3. $g^2 - 14g + 49$

4. $2w^2 - 4w + 9$

5. $4d^2 - 4d + 1$

6. $9n^2 + 30n + 25$

Factor each polynomial, if possible. If the polynomial cannot be factored, write *prime*.

7. $2x^2 - 72$

8. $6b^2 + 11b + 3$

9. $36t^2 - 24t + 4$

10. $4h^2 - 56$

11. $17a^2 - 24ab$

12. $q^2 - 14q + 36$

13. $y^2 + 24y + 144$

14. $6d^2 - 96$

Solve each equation. Check the solutions.

15. $x^2 - 18x + 81 = 0$

16. $4p^2 + 4p + 1 = 0$

17. $9g^2 - 12g + 4 = 0$

18. $y^2 - 16y + 64 = 81$

19. $4n^2 - 17 = 19$

20. $x^2 + 30x + 150 = -75$

21. $(k + 2)^2 = 16$

22. $(m - 4)^2 = 7$

8-6 Practice

Quadratic Equations: Perfect Squares

Determine whether each trinomial is a perfect square trinomial. Write *yes* or *no*. If so, factor it.

1. $m^2 + 16m + 64$

2. $9r^2 - 6r + 1$

3. $4y^2 - 20y + 25$

4. $16p^2 + 24p + 9$

5. $25b^2 - 4b + 16$

6. $49k^2 - 56k + 16$

Factor each polynomial, if possible. If the polynomial cannot be factored, write *prime*.

7. $3p^2 - 147$

8. $6x^2 + 11x - 35$

9. $50q^2 - 60q + 18$

10. $6t^3 - 14t^2 - 12t$

11. $6d^2 - 18$

12. $30k^2 + 38k + 12$

13. $15b^2 - 24bf$

14. $12h^2 - 60h + 75$

15. $9n^2 - 30n - 25$

16. $7u^2 - 28m^2$

17. $w^4 - 8w^2 - 9$

18. $16a^2 + 72ad + 81d^2$

Solve each equation. Check the solutions.

19. $4k^2 - 28k = -49$

20. $50b^2 + 20b + 2 = 0$

21. $\left(\frac{1}{2}t - 1\right)^2 = 0$

22. $g^2 + \frac{2}{3}g + \frac{1}{9} = 0$

23. $p^2 - \frac{6}{5}p + \frac{9}{25} = 0$

24. $x^2 + 12x + 36 = 25$

25. $y^2 - 8y + 16 = 64$

26. $(h + 9)^2 = 3$

27. $w^2 - 6w + 9 = 13$

28. GEOMETRY The area of a circle is given by the formula $A = \pi r^2$, where r is the radius. If increasing the radius of a circle by 1 inch gives the resulting circle an area of 100π square inches, what is the radius of the original circle?

29. PICTURE FRAMING Mikaela placed a frame around a print that measures 10 inches by 10 inches. The area of just the frame itself is 69 square inches. What is the width of the frame?

9-1 Skills Practice

Graphing Quadratic Functions

Use a table of values to graph each function. State the domain and the range.

1. $y = x^2 - 4$

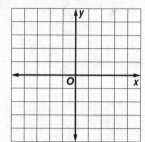

2. $y = -x^2 + 3$

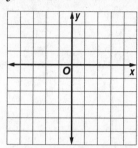

3. $y = x^2 - 2x - 6$

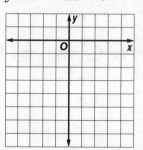

Find the vertex, the equation of the axis of symmetry, and the *y*-intercept.

4. $y = 2x^2 - 8x + 6$

5. $y = x^2 + 4x + 6$

6. $y = -3x^2 - 12x + 3$

Consider each equation.

a. Determine whether the function has a *maximum* or a *minimum* value.

b. State the maximum or minimum value.

c. What are the domain and range of the function?

7. $y = 2x^2$

8. $y = x^2 - 2x - 5$

9. $y = -x^2 + 4x - 1$

Graph each function.

10. $f(x) = -x^2 - 2x + 2$

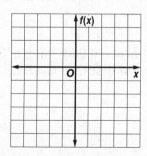

11. $f(x) = 2x^2 + 4x - 2$

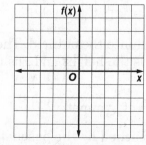

12. $f(x) = -2x^2 - 4x + 6$

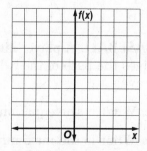

9-1 Practice

Graphing Quadratic Functions

Use a table of values to graph each function. Determine the domain and range.

1. $y = -x^2 + 2$

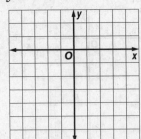

2. $y = x^2 - 6x + 3$

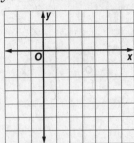

3. $y = -2x^2 - 8x - 5$

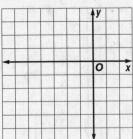

Find the vertex, the equation of the axis of symmetry, and the y-intercept.

4. $y = x^2 - 9$

5. $y = -2x^2 + 8x - 5$

6. $y = 4x^2 - 4x + 1$

Consider each equation. Determine whether the function has a *maximum* or a *minimum* value. State the maximum or minimum value. What are the domain and range of the function?

7. $y = 5x^2 - 2x + 2$

8. $y = -x^2 + 5x - 10$

9. $y = \frac{3}{2}x^2 + 4x - 9$

Graph each function.

10. $f(x) = -x^2 + 3$

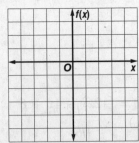

11. $f(x) = -2x^2 + 8x - 3$

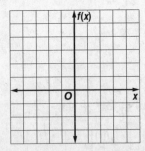

12. $f(x) = 2x^2 + 8x + 1$

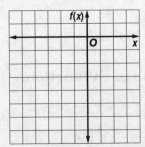

13. BASEBALL A player hits a baseball into the outfield. The equation $h = -0.005x^2 + x + 3$ gives the path of the ball, where h is the height and x is the horizontal distance the ball travels.

   a. What is the equation of the axis of symmetry?

   b. What is the maximum height reached by the baseball?

   c. An outfielder catches the ball three feet above the ground. How far has the ball traveled horizontally when the outfielder catches it?

9-2 Skills Practice

Solving Quadratic Equations by Graphing

Solve each equation by graphing.

1. $x^2 - 2x + 3 = 0$

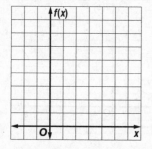

2. $c^2 + 6c + 8 = 0$

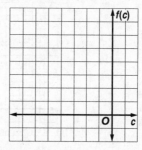

3. $a^2 - 2a = -1$

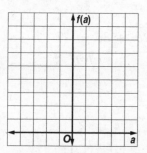

4. $n^2 - 7n = -10$

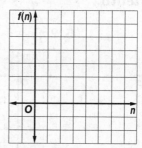

Solve each equation by graphing. If integral roots cannot be found, estimate the roots to the nearest tenth.

5. $p^2 + 4p + 2 = 0$

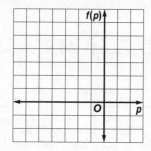

6. $x^2 + x - 3 = 0$

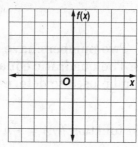

7. $d^2 + 6d = -3$

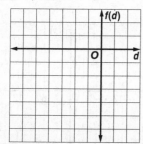

8. $h^2 + 1 = 4h$

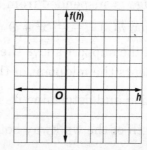

9-2 Practice

Solving Quadratic Equations by Graphing

Solve each equation by graphing.

1. $x^2 - 5x + 6 = 0$

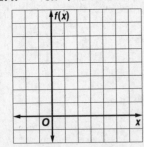

2. $w^2 + 6w + 9 = 0$

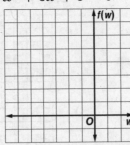

3. $b^2 - 3b + 4 = 0$

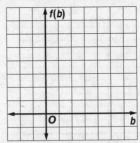

Solve each equation by graphing. If integral roots cannot be found, estimate the roots to the nearest tenth.

4. $p^2 + 4p = 3$

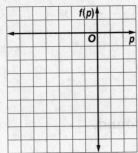

5. $2m^2 + 5 = 10m$

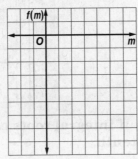

6. $2v^2 + 8v = -7$

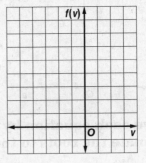

7. NUMBER THEORY Two numbers have a sum of 2 and a product of -8. The quadratic equation $-n^2 + 2n + 8 = 0$ can be used to determine the two numbers.

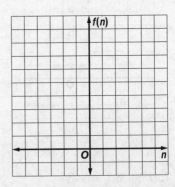

   a. Graph the related function $f(n) = -n^2 + 2n + 8$ and determine its x-intercepts.

   b. What are the two numbers?

8. DESIGN A footbridge is suspended from a parabolic support. The function $h(x) = -\frac{1}{25}x^2 + 9$ represents the height in feet of the support above the walkway, where $x = 0$ represents the midpoint of the bridge.

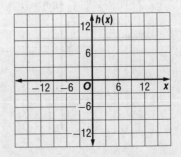

   a. Graph the function and determine its x-intercepts.

   b. What is the length of the walkway between the two supports?

9-3 Skills Practice

Transformations of Quadratic Functions

Describe how the graph of each function is related to the graph of $f(x) = x^2$.

1. $g(x) = x^2 + 2$

2. $h(x) = -1 + x^2$

3. $g(x) = x^2 - 8$

4. $h(x) = 7x^2$

5. $g(x) = \frac{1}{5}x^2$

6. $h(x) = -6x^2$

7. $g(x) = -x^2 + 3$

8. $h(x) = 5 - \frac{1}{2}x^2$

9. $g(x) = 4x^2 + 1$

Match each equation to its graph.

10. $y = 2x^2 - 2$

11. $y = \frac{1}{2}x^2 - 2$

12. $y = -\frac{1}{2}x^2 + 2$

13. $y = -2x^2 + 2$

A.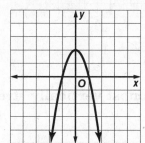

C.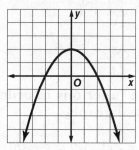

B.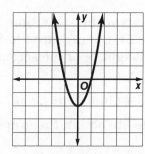

D.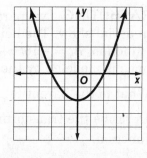

9-3 Practice

Transformations of Quadratic Functions

Describe how the graph of each function is related to the graph of $f(x) = x^2$.

1. $g(x) = 10 + x^2$

2. $h(x) = -\dfrac{2}{5} + x^2$

3. $g(x) = 9 - x^2$

4. $h(x) = 2x^2 + 2$

5. $g(x) = -\dfrac{3}{4}x^2 - \dfrac{1}{2}$

6. $h(x) = 4 - 3x^2$

Match each equation to its graph.

A.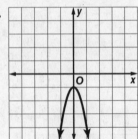

B.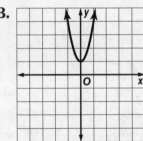

C.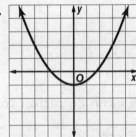

7. $y = -3x^2 - 1$

8. $y = \dfrac{1}{3}x^2 - 1$

9. $y = 3x^2 + 1$

List the functions in order from the most vertically stretched to the least vertically stretched graph.

10. $f(x) = 3x^2$, $g(x) = \dfrac{1}{2}x^2$, $h(x) = -2x^2$

11. $f(x) = \dfrac{1}{2}x^2$, $g(x) = -\dfrac{1}{6}x^2$, $h(x) = 4x^2$

12. PARACHUTING Two parachutists jump at the same time from two different planes as part of an aerial show. The height h_1 of the first parachutist in feet after t seconds is modeled by the function $h_1 = -16t^2 + 5000$. The height h_2 of the second parachutist in feet after t seconds is modeled by the function $h_2 = -16t^2 + 4000$.

a. What is the parent function of the two functions given?

b. Describe the transformations needed to obtain the graph of h_1 from the parent function.

c. Which parachutist will reach the ground first?

9-4 Skills Practice

Solving Quadratic Equations by Completing the Square

Find the value of c that makes each trinomial a perfect square.

1. $x^2 + 6x + c$

2. $x^2 + 4x + c$

3. $x^2 - 14x + c$

4. $x^2 - 2x + c$

5. $x^2 - 18x + c$

6. $x^2 + 20x + c$

7. $x^2 + 5x + c$

8. $x^2 - 70x + c$

9. $x^2 - 11x + c$

10. $x^2 + 9x + c$

Solve each equation by completing the square. Round to the nearest tenth if necessary.

11. $x^2 + 4x - 12 = 0$

12. $x^2 - 8x + 15 = 0$

13. $x^2 + 6x = 7$

14. $x^2 - 2x = 15$

15. $x^2 - 14x + 30 = 6$

16. $x^2 + 12x + 21 = 10$

17. $x^2 - 4x + 1 = 0$

18. $x^2 - 6x + 4 = 0$

19. $x^2 - 8x + 10 = 0$

20. $x^2 - 2x = 5$

21. $2x^2 + 20x = -2$

22. $0.5x^2 + 8x = -7$

9-4 Practice

Solving Quadratic Equations by Completing the Square

Find the value of c that makes each trinomial a perfect square.

1. $x^2 - 24x + c$

2. $x^2 + 28x + c$

3. $x^2 + 40x + c$

4. $x^2 + 3x + c$

5. $x^2 - 9x + c$

6. $x^2 - x + c$

Solve each equation by completing the square. Round to the nearest tenth if necessary.

7. $x^2 - 14x + 24 = 0$

8. $x^2 + 12x = 13$

9. $x^2 - 30x + 56 = -25$

10. $x^2 + 8x + 9 = 0$

11. $x^2 - 10x + 6 = -7$

12. $x^2 + 18x + 50 = 9$

13. $3x^2 + 15x - 3 = 0$

14. $4x^2 - 72 = 24x$

15. $0.9x^2 + 5.4x - 4 = 0$

16. $0.4x^2 + 0.8x = 0.2$

17. $\frac{1}{2}x^2 - x - 10 = 0$

18. $\frac{1}{4}x^2 + x - 2 = 0$

19. NUMBER THEORY The product of two consecutive even integers is 728. Find the integers.

20. BUSINESS Jaime owns a business making decorative boxes to store jewelry, mementos, and other valuables. The function $y = x^2 + 50x + 1800$ models the profit y that Jaime has made in month x for the first two years of his business.

a. Write an equation representing the month in which Jaime's profit is $2400.

b. Use completing the square to find out in which month Jaime's profit is $2400.

21. PHYSICS From a height of 256 feet above a lake on a cliff, Mikaela throws a rock out over the lake. The height H of the rock t seconds after Mikaela throws it is represented by the equation $H = -16t^2 + 32t + 256$. To the nearest tenth of a second, how long does it take the rock to reach the lake below? (*Hint:* Replace H with 0.)

9-5 Skills Practice

Solving Quadratic Equations by Using the Quadratic Formula

Solve each equation by using the Quadratic Formula. Round to the nearest tenth if necessary.

1. $x^2 - 49 = 0$

2. $x^2 - x - 20 = 0$

3. $x^2 - 5x - 36 = 0$

4. $x^2 + 11x + 30 = 0$

5. $x^2 - 7x = -3$

6. $x^2 + 4x = -1$

7. $x^2 - 9x + 22 = 0$

8. $x^2 + 6x + 3 = 0$

9. $2x^2 + 5x - 7 = 0$

10. $2x^2 - 3x = -1$

11. $2x^2 + 5x + 4 = 0$

12. $2x^2 + 7x = 9$

13. $3x^2 + 2x - 3 = 0$

14. $3x^2 - 7x - 6 = 0$

State the value of the discriminant for each equation. Then determine the number of real solutions of the equation.

15. $x^2 + 4x + 3 = 0$

16. $x^2 + 2x + 1 = 0$

17. $x^2 - 4x + 10 = 0$

18. $x^2 - 6x + 7 = 0$

19. $x^2 - 2x - 7 = 0$

20. $x^2 - 10x + 25 = 0$

21. $2x^2 + 5x - 8 = 0$

22. $2x^2 + 6x + 12 = 0$

23. $2x^2 - 4x + 10 = 0$

24. $3x^2 + 7x + 3 = 0$

9-5　Practice

Solving Quadratic Equations by Using the Quadratic Formula

Solve each equation by using the Quadratic Formula. Round to the nearest tenth if necessary.

1. $x^2 + 2x - 3 = 0$

2. $x^2 + 8x + 7 = 0$

3. $x^2 - 4x + 6 = 0$

4. $x^2 - 6x + 7 = 0$

5. $2x^2 + 9x - 5 = 0$

6. $2x^2 + 12x + 10 = 0$

7. $2x^2 - 9x = -12$

8. $2x^2 - 5x = 12$

9. $3x^2 + x = 4$

10. $3x^2 - 1 = -8x$

11. $4x^2 + 7x = 15$

12. $1.6x^2 + 2x + 2.5 = 0$

13. $4.5x^2 + 4x - 1.5 = 0$

14. $\frac{1}{2}x^2 + 2x + \frac{3}{2} = 0$

15. $3x^2 - \frac{3}{4}x = \frac{1}{2}$

State the value of the discriminant for each equation. Then determine the number of real solutions of the equation.

16. $x^2 + 8x + 16 = 0$

17. $x^2 + 3x + 12 = 0$

18. $2x^2 + 12x = -7$

19. $2x^2 + 15x = -30$

20. $4x^2 + 9 = 12x$

21. $3x^2 - 2x = 3.5$

22. $2.5x^2 + 3x - 0.5 = 0$

23. $\frac{3}{4}x^2 - 3x = -4$

24. $\frac{1}{4}x^2 = -x - 1$

25. **CONSTRUCTION** A roofer tosses a piece of roofing tile from a roof onto the ground 30 feet below. He tosses the tile with an initial downward velocity of 10 feet per second.

   a. Write an equation to find how long it takes the tile to hit the ground. Use the model for vertical motion, $H = -16t^2 + vt + h$, where H is the height of an object after t seconds, v is the initial velocity, and h is the initial height. (*Hint*: Since the object is thrown down, the initial velocity is negative.)

   b. How long does it take the tile to hit the ground?

26. **PHYSICS** Lupe tosses a ball up to Quyen, waiting at a third-story window, with an initial velocity of 30 feet per second. She releases the ball from a height of 6 feet. The equation $h = -16t^2 + 30t + 6$ represents the height h of the ball after t seconds. If the ball must reach a height of 25 feet for Quyen to catch it, does the ball reach Quyen? Explain. (*Hint:* Substitute 25 for h and use the discriminant.)

9-6 | Skills Practice

Exponential Functions

Graph each function. Find the *y*-intercept, and state the domain and range. Then use the graph to determine the approximate value of the given expression. Use a calculator to confirm the value.

1. $y = 2^x$; $2^{2.3}$

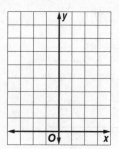

2. $y = \left(\frac{1}{3}\right)^x$; $\left(\frac{1}{3}\right)^{-1.6}$

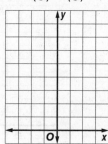

Graph each function. Find the *y*-intercept, and state the domain and range.

3. $y = 3(2^x)$

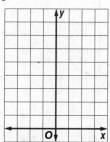

4. $y = 3^x + 2$

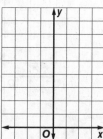

Determine whether the set of data shown below displays exponential behavior. Write *yes* or *no*. Explain why or why not.

5.

x	−3	−2	−1	0
y	9	12	15	18

6.

x	0	5	10	15
y	20	10	5	2.5

7.

x	4	8	12	16
y	20	40	80	160

8.

x	50	30	10	−10
y	90	70	50	30

9-6 Practice

Exponential Functions

Graph each function. Find the *y*-intercept and state the domain and range. Then use the graph to determine the approximate value of the given expression. Use a calculator to confirm the value.

1. $y = \left(\frac{1}{10}\right)^x; \left(\frac{1}{10}\right)^{-0.5}$

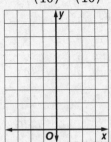

2. $y = 3^x; 3^{1.9}$

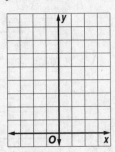

3. $y = \left(\frac{1}{4}\right)^x; \left(\frac{1}{4}\right)^{-1.4}$

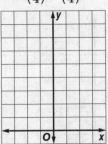

Graph each function. Find the *y*-intercept, and state the domain and range.

4. $y = 4(2^x) + 1$

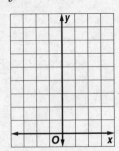

5. $y = 2(2^x - 1)$

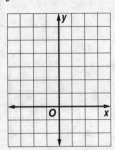

6. $y = 0.5(3^x - 3)$

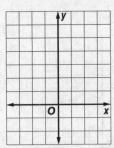

Determine whether the set of data shown below displays exponential behavior. Write *yes* or *no*. Explain why or why not.

7.

x	2	5	8	11
y	480	120	30	7.5

8.

x	21	18	15	12
y	30	23	16	9

9. LEARNING Ms. Klemperer told her English class that each week students tend to forget one sixth of the vocabulary words they learned the previous week. Suppose a student learns 60 words. The number of words remembered can be described by the function $W(x) = 60\left(\frac{5}{6}\right)^x$, where *x* is the number of weeks that pass. How many words will the student remember after 3 weeks?

10. BIOLOGY Suppose a certain cell reproduces itself in four hours. If a lab researcher begins with 50 cells, how many cells will there be after one day, two days, and three days? (*Hint:* Use the exponential function $y = 50(2^x)$.)

9-7　Skills Practice

Growth and Decay

1. POPULATION The population of New York City increased from 8,008,278 in 2000 to 8,168,388 in 2005. The annual rate of population increase for the period was about 0.4%.

a. Write an equation for the population t years after 2000.

b. Use the equation to predict the population of New York City in 2015.

2. SAVINGS The Fresh and Green Company has a savings plan for its employees. If an employee makes an initial contribution of $1000, the company pays 8% interest compounded quarterly.

a. If an employee participating in the plan withdraws the balance of the account after 5 years, how much will be in the account?

b. If an employee participating in the plan withdraws the balance of the account after 35 years, how much will be in the account?

3. HOUSING Mr. and Mrs. Boyce bought a house for $96,000 in 1995. The real estate broker indicated that houses in their area were appreciating at an average annual rate of 7%. If the appreciation remained steady at this rate, what was the value of the Boyce's home in 2009?

4. MANUFACTURING Zeller Industries bought a piece of weaving equipment for $60,000. It is expected to depreciate at an average rate of 10% per year.

a. Write an equation for the value of the piece of equipment after t years.

b. Find the value of the piece of equipment after 6 years.

5. FINANCES Kyle saved $500 from a summer job. He plans to spend 10% of his savings each week on various forms of entertainment. At this rate, how much will Kyle have left after 15 weeks?

6. TRANSPORTATION Tiffany's mother bought a car for $9000 five years ago. She wants to sell it to Tiffany based on a 15% annual rate of depreciation. At this rate, how much will Tiffany pay for the car?

9-7 Practice

Growth and Decay

1. **COMMUNICATIONS** Sports radio stations numbered 220 in 1996. The number of sports radio stations has since increased by approximately 14.3% per year.

   a. Write an equation for the number of sports radio stations for t years after 1996.

   b. If the trend continues, predict the number of sports radio stations in this format for the year 2010.

2. **INVESTMENTS** Determine the amount of an investment if $500 is invested at an interest rate of 4.25% compounded quarterly for 12 years.

3. **INVESTMENTS** Determine the amount of an investment if $300 is invested at an interest rate of 6.75% compounded semiannually for 20 years.

4. **HOUSING** The Greens bought a condominium for $110,000 in 2005. If its value appreciates at an average rate of 6% per year, what will the value be in 2010?

5. **DEFORESTATION** During the 1990s, the forested area of Guatemala decreased at an average rate of 1.7%.

   a. If the forested area in Guatemala in 1990 was about 34,400 square kilometers, write an equation for the forested area for t years after 1990.

   b. If this trend continues, predict the forested area in 2015.

6. **BUSINESS** A piece of machinery valued at $25,000 depreciates at a steady rate of 10% yearly. What will the value of the piece of machinery be after 7 years?

7. **TRANSPORTATION** A new car costs $18,000. It is expected to depreciate at an average rate of 12% per year. Find the value of the car in 8 years.

8. **POPULATION** The population of Osaka, Japan, declined at an average annual rate of 0.05% for the five years between 1995 and 2000. If the population of Osaka was 11,013,000 in 2000 and it continues to decline at the same rate, predict the population in 2050.

9-8 Skills Practice

Geometric Sequences as Exponential Functions

Determine whether each sequence is *arithmetic*, *geometric*, or *neither*. Explain.

1. 7, 13, 19, 25, …

2. −96, −48, −24, −12, …

3. 108, 66, 141, 99, …

4. 3, 9, 81, 6561, …

5. $\frac{7}{3}$, 14, 84, 504, …

6. $\frac{3}{8}$, $-\frac{1}{8}$, $-\frac{5}{8}$, $-\frac{9}{8}$, …

Find the next three terms in each geometric sequence.

7. 2500, 500, 100, …

8. 2, 6, 18, …

9. −4, 24, −144, …

10. $\frac{4}{5}$, $\frac{2}{5}$, $\frac{1}{5}$, …

11. −3, −12, −48, …

12. 72, 12, 2, …

13. Write an equation for the nth term of the geometric sequence 3, −24, 192, ….
Find the ninth term of this sequence.

14. Write an equation for the nth term of the geometric sequence $\frac{9}{16}$, $\frac{3}{8}$, $\frac{1}{4}$, ….
Find the seventh term of this sequence.

15. Write an equation for the nth term of the geometric sequence 1000, 200, 40, ….
Find the fifth term of this sequence.

16. Write an equation for the nth term of the geometric sequence −8, −2, −$\frac{1}{2}$, ….
Find the eighth term of this sequence.

17. Write an equation for the nth term of the geometric sequence 32, 48, 72, ….
Find the sixth term of this sequence.

18. Write an equation for the nth term of the geometric sequence $\frac{3}{100}$, $\frac{3}{10}$, 3, ….
Find the ninth term of this sequence.

9-8 Practice

Geometric Sequences as Exponential Functions

Determine whether each sequence is *arithmetic*, *geometric*, or *neither*. Explain.

1. 1, −5, −11, −17, …

2. 3, $\frac{3}{2}$, 1, $\frac{3}{4}$, …

3. 108, 36, 12, 4, …

4. −2, 4, −6, 8, …

Find the next three terms in each geometric sequence.

5. 64, 16, 4, …

6. 2, −12, 72, …

7. 3750, 750, 150, …

8. 4, 28, 196, …

9. Write an equation for the nth term of the geometric sequence 896, −448, 224, … . Find the eighth term of this sequence.

10. Write an equation for the nth term of the geometric sequence 3584, 896, 224, … . Find the sixth term of this sequence.

11. Find the sixth term of a geometric sequence for which $a_2 = 288$ and $r = \frac{1}{4}$.

12. Find the eighth term of a geometric sequence for which $a_3 = 35$ and $r = 7$.

13. PENNIES Thomas is saving pennies in a jar. The first day he saves 3 pennies, the second day 12 pennies, the third day 48 pennies, and so on. How many pennies does Thomas save on the eighth day?

9-9 Skills Practice

Analyzing Functions with Successive Differences and Ratios

Graph each set of ordered pairs. Determine whether the ordered pairs represent a *linear* function, a *quadratic* function, or an *exponential* function.

1. (2, 3), (1, 1), (0, –1), (–1, –3), (–3, –5)

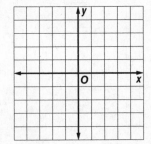

2. (–1, 0.5), (0, 1), (1, 2), (2, 4)

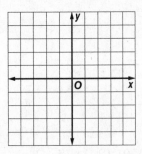

3. (–2, 4), (–1, 1), (0, 0), (1, 1), (2, 4)

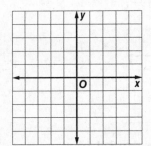

4. (–3, 5), (–2, 2), (–1, 1), (0, 2), (1, 5)

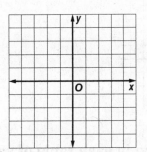

Look for a pattern in each table of values to determine which model best describes the data. Then write an equation for the function that models the data.

5.

x	–3	–2	–1	0	1	2
y	32	16	8	4	2	1

6.

x	–1	0	1	2	3
y	7	3	–1	–5	–9

7.

x	–3	–2	–1	0	1
y	–27	–12	–3	0	–3

8.

x	0	1	2	3	4
y	0.5	1.5	4.5	13.5	40.5

9.

x	–2	–1	0	1	2
y	–8	–4	0	4	8

9-9 Practice

Analyzing Functions with Successive Differences and Ratios

Graph each set of ordered pairs. Determine whether the ordered pairs represent a *linear* function, a *quadratic* function, or an *exponential* function.

1. (4, 0.5), (3, 1.5), (2, 2.5), (1, 3.5), (0, 4.5)

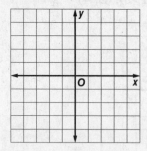

2. $\left(-1, \dfrac{1}{9}\right)$, $\left(0, \dfrac{1}{3}\right)$, (1, 1), (2, 3)

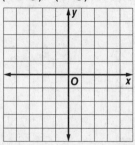

3. (−4, 4), (−2, 1), (0, 0), (2, 1), (4, 4)

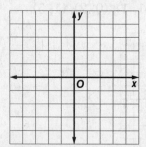

4. (−4, 2), (−2, 1), (0, 0), (2, −1), (4, −2)

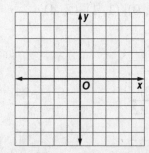

Look for a pattern in each table of values to determine which model best describes the data. Then write an equation for the function that models the data.

5.

x	−3	−1	1	3	5
y	−5	−2	1	4	7

6.

x	−2	−1	0	1	2
y	0.02	0.2	2	20	200

7.

x	−1	0	1	2	3
y	6	0	6	24	54

8.

x	−2	−1	0	1	2
y	18	9	0	−9	−18

9. INSECTS The local zoo keeps track of the number of dragonflies breeding in their insect exhibit each day.

Day	1	2	3	4	5
Dragonflies	9	18	36	72	144

a. Determine which function best models the data.

b. Write an equation for the function that models the data.

c. Use your equation to determine the number of dragonflies that will be breeding after 9 days.

10-1 Skills Practice

Square Root Functions

Graph each function, and compare to the parent graph. State the domain and range.

1. $y = 2\sqrt{x}$

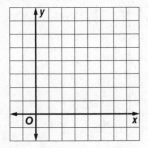

2. $y = \frac{1}{2}\sqrt{x}$

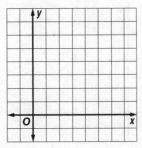

3. $y = 5\sqrt{x}$

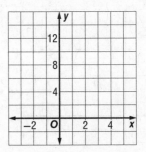

4. $y = \sqrt{x} + 1$

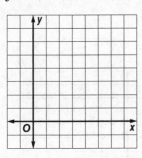

5. $y = \sqrt{x} - 4$

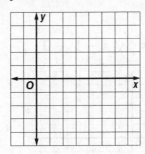

6. $y = \sqrt{x - 1}$

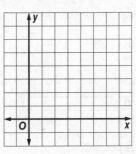

7. $y = -\sqrt{x - 3}$

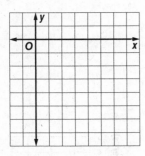

8. $y = \sqrt{x - 2} + 3$

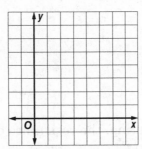

9. $y = -\frac{1}{2}\sqrt{x - 4} + 1$

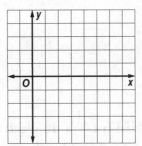

10-1 Practice

Square Root Functions

Graph each function, and compare to the parent graph. State the domain and range.

1. $y = \frac{4}{3}\sqrt{x}$

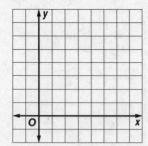

2. $y = \sqrt{x} + 2$

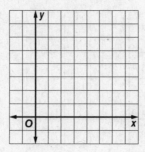

3. $y = \sqrt{x - 3}$

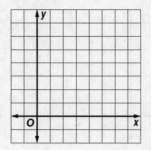

4. $y = -\sqrt{x} + 1$

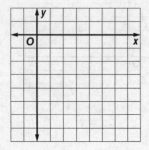

5. $y = 2\sqrt{x - 1} + 1$

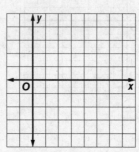

6. $y = -\sqrt{x - 2} + 2$

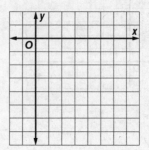

7. OHM'S LAW In electrical engineering, the resistance of a circuit can be found by the equation $I = \sqrt{\frac{P}{R}}$, where I is the current in amperes, P is the power in watts, and R is the resistance of the circuit in ohms. Graph this function for a circuit with a resistance of 4 ohms.

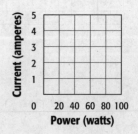

10-2 Skills Practice

Simplifying Radical Expressions

Simplify each expression.

1. $\sqrt{28}$

2. $\sqrt{40}$

3. $\sqrt{72}$

4. $\sqrt{99}$

5. $\sqrt{2} \cdot \sqrt{10}$

6. $\sqrt{5} \cdot \sqrt{60}$

7. $3\sqrt{5} \cdot \sqrt{5}$

8. $\sqrt{6} \cdot 4\sqrt{24}$

9. $2\sqrt{3} \cdot 3\sqrt{15}$

10. $\sqrt{16b^4}$

11. $\sqrt{81a^2d^4}$

12. $\sqrt{40x^4y^6}$

13. $\sqrt{75m^5p^2}$

14. $\sqrt{\dfrac{5}{3}}$

15. $\sqrt{\dfrac{1}{6}}$

16. $\sqrt{\dfrac{6}{7}} \cdot \sqrt{\dfrac{1}{3}}$

17. $\sqrt{\dfrac{q}{12}}$

18. $\sqrt{\dfrac{4h}{5}}$

19. $\sqrt{\dfrac{12}{b^2}}$

20. $\sqrt{\dfrac{45}{4m^4}}$

21. $\dfrac{2}{4 + \sqrt{5}}$

22. $\dfrac{3}{2 - \sqrt{3}}$

23. $\dfrac{5}{7 + \sqrt{7}}$

24. $\dfrac{4}{3 - \sqrt{2}}$

10-2 **Practice**

Simplifying Radical Expressions

Simplify.

1. $\sqrt{24}$

2. $\sqrt{60}$

3. $\sqrt{108}$

4. $\sqrt{8} \cdot \sqrt{6}$

5. $\sqrt{7} \cdot \sqrt{14}$

6. $3\sqrt{12} \cdot 5\sqrt{6}$

7. $4\sqrt{3} \cdot 3\sqrt{18}$

8. $\sqrt{27tu^3}$

9. $\sqrt{50p^5}$

10. $\sqrt{108x^6y^4z^5}$

11. $\sqrt{56m^2n^4p^5}$

12. $\dfrac{\sqrt{8}}{\sqrt{6}}$

13. $\sqrt{\dfrac{2}{10}}$

14. $\sqrt{\dfrac{5}{32}}$

15. $\sqrt{\dfrac{3}{4}} \cdot \sqrt{\dfrac{4}{5}}$

16. $\sqrt{\dfrac{1}{7}} \cdot \sqrt{\dfrac{7}{11}}$

17. $\dfrac{\sqrt{3k}}{\sqrt{8}}$

18. $\sqrt{\dfrac{18}{x^3}}$

19. $\sqrt{\dfrac{4y}{3y^2}}$

20. $\sqrt{\dfrac{9ab}{4ab^4}}$

21. $\dfrac{3}{5 - \sqrt{2}}$

22. $\dfrac{8}{3 + \sqrt{3}}$

23. $\dfrac{5}{\sqrt{7} + \sqrt{3}}$

24. $\dfrac{3\sqrt{7}}{-1 - \sqrt{27}}$

25. **SKYDIVING** When a skydiver jumps from an airplane, the time t it takes to free fall a given distance can be estimated by the formula $t = \sqrt{\dfrac{2s}{9.8}}$, where t is in seconds and s is in meters. If Julie jumps from an airplane, how long will it take her to free fall 750 meters?

26. **METEOROLOGY** To estimate how long a thunderstorm will last, meteorologists can use the formula $t = \sqrt{\dfrac{d^3}{216}}$, where t is the time in hours and d is the diameter of the storm in miles.

    a. A thunderstorm is 8 miles in diameter. Estimate how long the storm will last. Give your answer in simplified form and as a decimal.

    b. Will a thunderstorm twice this diameter last twice as long? Explain.

10-3 Skills Practice

Operations with Radical Expressions

Simplify each expression.

1. $7\sqrt{7} - 2\sqrt{7}$

2. $3\sqrt{13} + 7\sqrt{13}$

3. $6\sqrt{5} - 2\sqrt{5} + 8\sqrt{5}$

4. $\sqrt{15} + 8\sqrt{15} - 12\sqrt{15}$

5. $12\sqrt{r} - 9\sqrt{r}$

6. $9\sqrt{6a} - 11\sqrt{6a} + 4\sqrt{6a}$

7. $\sqrt{44} - \sqrt{11}$

8. $\sqrt{28} + \sqrt{63}$

9. $4\sqrt{3} + 2\sqrt{12}$

10. $8\sqrt{54} - 4\sqrt{6}$

11. $\sqrt{27} + \sqrt{48} + \sqrt{12}$

12. $\sqrt{72} + \sqrt{50} - \sqrt{8}$

13. $\sqrt{180} - 5\sqrt{5} + \sqrt{20}$

14. $2\sqrt{24} + 4\sqrt{54} + 5\sqrt{96}$

15. $5\sqrt{8} + 2\sqrt{20} - \sqrt{8}$

16. $2\sqrt{13} + 4\sqrt{2} - 5\sqrt{13} + \sqrt{2}$

17. $\sqrt{2}\left(\sqrt{8} + \sqrt{6}\right)$

18. $\sqrt{5}\left(\sqrt{10} - \sqrt{3}\right)$

19. $\sqrt{6}\left(3\sqrt{2} - 2\sqrt{3}\right)$

20. $3\sqrt{3}\left(2\sqrt{6} + 4\sqrt{10}\right)$

21. $\left(4 + \sqrt{3}\right)\left(4 - \sqrt{3}\right)$

22. $\left(2 - \sqrt{6}\right)^2$

23. $\left(\sqrt{8} + \sqrt{2}\right)\left(\sqrt{5} + \sqrt{3}\right)$

24. $\left(\sqrt{6} + 4\sqrt{5}\right)\left(4\sqrt{3} - \sqrt{10}\right)$

10-3 Practice

Operations with Radical Expressions

Simplify each expression.

1. $8\sqrt{30} - 4\sqrt{30}$

2. $2\sqrt{5} - 7\sqrt{5} - 5\sqrt{5}$

3. $7\sqrt{13x} - 14\sqrt{13x} + 2\sqrt{13x}$

4. $2\sqrt{45} + 4\sqrt{20}$

5. $\sqrt{40} - \sqrt{10} + \sqrt{90}$

6. $2\sqrt{32} + 3\sqrt{50} - 3\sqrt{18}$

7. $\sqrt{27} + \sqrt{18} + \sqrt{300}$

8. $5\sqrt{8} + 3\sqrt{20} - \sqrt{32}$

9. $\sqrt{14} - \sqrt{\dfrac{2}{7}}$

10. $\sqrt{50} + \sqrt{32} - \sqrt{\dfrac{1}{2}}$

11. $5\sqrt{19} + 4\sqrt{28} - 8\sqrt{19} + \sqrt{63}$

12. $3\sqrt{10} + \sqrt{75} - 2\sqrt{40} - 4\sqrt{12}$

13. $\sqrt{6}\left(\sqrt{10} + \sqrt{15}\right)$

14. $\sqrt{5}\left(5\sqrt{2} - 4\sqrt{8}\right)$

15. $2\sqrt{7}\left(3\sqrt{12} + 5\sqrt{8}\right)$

16. $\left(5 - \sqrt{15}\right)^2$

17. $\left(\sqrt{10} + \sqrt{6}\right)\left(\sqrt{30} - \sqrt{18}\right)$

18. $\left(\sqrt{8} + \sqrt{12}\right)\left(\sqrt{48} + \sqrt{18}\right)$

19. $\left(\sqrt{2} + 2\sqrt{8}\right)\left(3\sqrt{6} - \sqrt{5}\right)$

20. $\left(4\sqrt{3} - 2\sqrt{5}\right)\left(3\sqrt{10} + 5\sqrt{6}\right)$

21. SOUND The speed of sound V in meters per second near Earth's surface is given by $V = 20\sqrt{t + 273}$, where t is the surface temperature in degrees Celsius.

  a. What is the speed of sound near Earth's surface at 15°C and at 2°C in simplest form?

  b. How much faster is the speed of sound at 15°C than at 2°C?

22. GEOMETRY A rectangle is $5\sqrt{7} + 2\sqrt{3}$ meters long and $6\sqrt{7} - 3\sqrt{3}$ meters wide.

  a. Find the perimeter of the rectangle in simplest form.

  b. Find the area of the rectangle in simplest form.

10-4 Skills Practice

Radical Equations

Solve each equation. Check your solution.

1. $\sqrt{f} = 7$

2. $\sqrt{-x} = 5$

3. $\sqrt{5p} = 10$

4. $\sqrt{4y} = 6$

5. $2\sqrt{2} = \sqrt{u}$

6. $3\sqrt{5} = \sqrt{-n}$

7. $\sqrt{g} - 6 = 3$

8. $\sqrt{5a} + 2 = 0$

9. $\sqrt{2t - 1} = 5$

10. $\sqrt{3k - 2} = 4$

11. $\sqrt{x + 4} - 2 = 1$

12. $\sqrt{4x - 4} - 4 = 0$

13. $\dfrac{\sqrt{d}}{3} = 4$

14. $\sqrt{\dfrac{m}{3}} = 3$

15. $x = \sqrt{x + 2}$

16. $d = \sqrt{12 - d}$

17. $\sqrt{6x - 9} = x$

18. $\sqrt{6p - 8} = p$

19. $\sqrt{x + 5} = x - 1$

20. $\sqrt{8 - d} = d - 8$

21. $\sqrt{r - 3} + 5 = r$

22. $\sqrt{y - 1} + 3 = y$

23. $\sqrt{5n + 4} = n + 2$

24. $\sqrt{3z - 6} = z - 2$

10-4 Practice

Radical Equations

Solve each equation. Check your solution.

1. $\sqrt{-b} = 8$

2. $4\sqrt{3} = \sqrt{x}$

3. $2\sqrt{4r} + 3 = 11$

4. $6 - \sqrt{2y} = -2$

5. $\sqrt{k + 2} - 3 = 7$

6. $\sqrt{m - 5} = 4\sqrt{3}$

7. $\sqrt{6t + 12} = 8\sqrt{6}$

8. $\sqrt{3j - 11} + 2 = 9$

9. $\sqrt{2x + 15} + 5 = 18$

10. $\sqrt{\dfrac{3d}{5}} - 4 = 2$

11. $6\sqrt{\dfrac{3x}{3}} - 3 = 0$

12. $6 + \sqrt{\dfrac{5r}{6}} = -2$

13. $y = \sqrt{y + 6}$

14. $\sqrt{15 - 2x} = x$

15. $\sqrt{w + 4} = w + 4$

16. $\sqrt{17 - k} = k - 5$

17. $\sqrt{5m - 16} = m - 2$

18. $\sqrt{24 + 8q} = q + 3$

19. $\sqrt{4t + 17} - t - 3 = 0$

20. $4 - \sqrt{3m + 28} = m$

21. $\sqrt{10p + 61} - 7 = p$

22. $\sqrt{2x^2 - 9} = x$

23. ELECTRICITY The voltage V in a circuit is given by $V = \sqrt{PR}$, where P is the power in watts and R is the resistance in ohms.

   a. If the voltage in a circuit is 120 volts and the circuit produces 1500 watts of power, what is the resistance in the circuit?

   b. Suppose an electrician designs a circuit with 110 volts and a resistance of 10 ohms. How much power will the circuit produce?

24. FREE FALL Assuming no air resistance, the time t in seconds that it takes an object to fall h feet can be determined by the equation $t = \dfrac{\sqrt{h}}{4}$.

   a. If a skydiver jumps from an airplane and free falls for 10 seconds before opening the parachute, how many feet does the skydiver fall?

   b. Suppose a second skydiver jumps and free falls for 6 seconds. How many feet does the second skydiver fall?

10-5 Skills Practice

The Pythagorean Theorem

Find each missing length. If necessary, round to the nearest hundredth.

1.
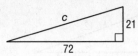

2.
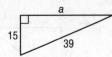

3.
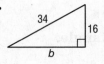

4.

5.
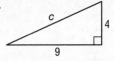

6.
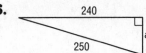

Determine whether each set of measures can be sides of a right triangle. Then determine whether they form a Pythagorean triple.

7. 7, 24, 25

8. 15, 30, 34

9. 16, 28, 32

10. 18, 24, 30

11. 15, 36, 39

12. 5, 7, $\sqrt{74}$

13. 4, 5, 6

14. 10, 11, $\sqrt{221}$

10-5 Practice

The Pythagorean Theorem

Find each missing length. If necessary, round to the nearest hundredth.

1.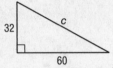

2.

3.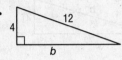

Determine whether each set of measures can be sides of a right triangle. Then determine whether they form a Pythagorean triple.

4. 11, 18, 21

5. 21, 72, 75

6. 7, 8, 11

7. 9, 10, $\sqrt{161}$

8. 9, $2\sqrt{10}$, 11

9. $\sqrt{7}$, $2\sqrt{2}$, $\sqrt{15}$

10. **STORAGE** The shed in Stephan's back yard has a door that measures 6 feet high and 3 feet wide. Stephan would like to store a square theater prop that is 7 feet on a side. Will it fit through the door diagonally? Explain.

11. **SCREEN SIZES** The size of a television is measured by the length of the screen's diagonal.

  a. If a television screen measures 24 inches high and 18 inches wide, what size television is it?

  b. Darla told Tri that she has a 35-inch television. The height of the screen is 21 inches. What is its width?

  c. Tri told Darla that he has a 5-inch handheld television and that the screen measures 2 inches by 3 inches. Is this a reasonable measure for the screen size? Explain.

10-6 Skills Practice

The Distance and Midpoint Formulas

Find the distance between the points with the given coordinates.

1. $(9, 7), (1, 1)$

2. $(5, 2), (8, -2)$

3. $(1, -3), (1, 4)$

4. $(7, 2), (-5, 7)$

5. $(-6, 3), (10, 3)$

6. $(3, 3), (-2, 3)$

7. $(-1, -4), (-6, 0)$

8. $(-2, 4), (5, 8)$

Find the possible values of a if the points with the given coordinates are the indicated distance apart.

9. $(-2, -5), (a, 7); d = 13$

10. $(8, -2), (5, a); d = 3$

11. $(4, a), (1, 6); d = 5$

12. $(a, 3), (5, -1); d = 5$

13. $(1, 1), (a, 1); d = 4$

14. $(2, a), (2, 3); d = 10$

15. $(a, 2), (-3, 3); d = \sqrt{2}$

16. $(-5, 3), (-3, a); d = \sqrt{5}$

Find the coordinates of the midpoint of the segment with the given endpoints.

17. $(-3, 4), (-2, 8)$

18. $(5, -6), (7, -9)$

19. $(4, 2), (8, 6)$

20. $(5, 2), (3, 10)$

21. $(12, -1), (4, -11)$

22. $(-3, -1), (-11, 3)$

23. $(9, 3), (6, -6)$

24. $(0, -4), (8, 4)$

10-6 Practice

The Distance and Midpoint Formulas

Find the distance between the points with the given coordinates.

1. $(4, 7), (1, 3)$

2. $(0, 9), (-7, -2)$

3. $(6, 2), \left(4, \frac{1}{2}\right)$

4. $(-1, 7), \left(\frac{1}{3}, 6\right)$

5. $(\sqrt{3}, 3), (2\sqrt{3}, 5)$

6. $(2\sqrt{2}, -1), (3\sqrt{2}, 3)$

Find the possible values of a if the points with the given coordinates are the indicated distance apart.

7. $(4, -1), (a, 5); d = 10$

8. $(2, -5), (a, 7); d = 15$

9. $(6, -7), (a, -4); d = \sqrt{18}$

10. $(-4, 1), (a, 8); d = \sqrt{50}$

11. $(8, -5), (a, 4); d = \sqrt{85}$

12. $(-9, 7), (a, 5); d = \sqrt{29}$

Find the coordinates of the midpoint of the segment with the given endpoints.

13. $(4, -6), (3, -9)$

14. $(-3, -8), (-7, 2)$

15. $(0, -4), (3, 2)$

16. $(-13, -9), (-1, -5)$

17. $\left(2, -\frac{1}{2}\right), \left(1, \frac{1}{2}\right)$

18. $\left(\frac{2}{3}, -1\right), \left(2, \frac{1}{3}\right)$

19. **BASEBALL** Three players are warming up for a baseball game. Player B stands 9 feet to the right and 18 feet in front of Player A. Player C stands 8 feet to the left and 13 feet in front of Player A.

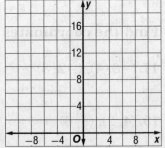

  a. Draw a model of the situation on the coordinate grid. Assume that Player A is located at $(0, 0)$.

  b. To the nearest tenth, what is the distance between Players A and B and between Players A and C?

  c. What is the distance between Players B and C?

20. **MAPS** Maria and Jackson live in adjacent neighborhoods. If they superimpose a coordinate grid on the map of their neighborhoods, Maria lives at $(-9, 1)$ and Jackson lives at $(5, -4)$. If each unit on the grid represents $\frac{1}{4}$ mile, how far apart are Maria's and Jackson's houses?

10-7 Skills Practice

Similar Triangles

Determine whether each pair of triangles is similar. Justify your answer.

1.
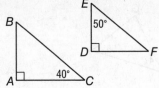

2.
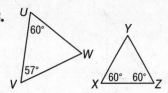

3.
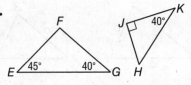

4.
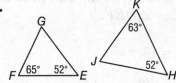

Find the missing measures for the pair of similar triangles if $\triangle PQR \sim \triangle STU$.

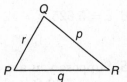

5. $r = 4, s = 6, t = 3, u = 2$

6. $t = 8, p = 21, q = 14, r = 7$

7. $p = 15, q = 10, r = 5, s = 6$

8. $p = 48, s = 16, t = 8, u = 4$

9. $q = 6, s = 2, t = \dfrac{3}{2}, u = \dfrac{1}{2}$

10. $p = 3, q = 2, r = 1, u = \dfrac{1}{3}$

11. $p = 14, q = 7, u = 2.5, t = 5$

12. $r = 6, s = 3, t = \dfrac{21}{8}, u = \dfrac{9}{4}$

10-7 Practice

Similar Triangles

Determine whether each pair of triangles is similar. Justify your answer.

1.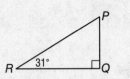

2.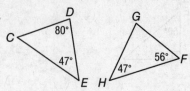

Find the missing measures for the pair of similar triangles if $\triangle ABC \sim \triangle DEF$.

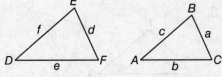

3. $c = 4$, $d = 12$, $e = 16$, $f = 8$

4. $e = 20$, $a = 24$, $b = 30$, $c = 15$

5. $a = 10$, $b = 12$, $c = 6$, $d = 4$

6. $a = 4$, $d = 6$, $e = 4$, $f = 3$

7. $b = 15$, $d = 16$, $e = 20$, $f = 10$

8. $a = 16$, $b = 22$, $c = 12$, $f = 8$

9. $a = \frac{5}{2}$, $b = 3$, $f = \frac{11}{2}$, $e = 7$

10. $c = 4$, $d = 6$, $e = 5.625$, $f = 12$

11. SHADOWS Suppose you are standing near a building and you want to know its height. The building casts a 66-foot shadow. You cast a 3-foot shadow. If you are 5 feet 6 inches tall, how tall is the building?

12. MODELS Truss bridges use triangles in their support beams. Molly made a model of a truss bridge in the scale of 1 inch = 8 feet. If the height of the triangles on the model is 4.5 inches, what is the height of the triangles on the actual bridge?

10-8 Skills Practice

Trigonometric Ratios

Find the values of the three trigonometric ratios for angle *A*.

1.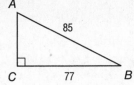

2.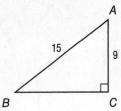

3.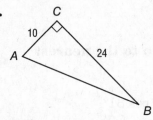

4.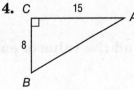

Use a calculator to find the value of each trigonometric ratio to the nearest ten-thousandth.

5. sin 18°   6. cos 68°   7. tan 27°

8. cos 60°   9. tan 75°   10. sin 9°

Solve each right triangle. Round each side length to the nearest tenth.

11.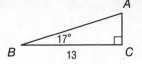

12.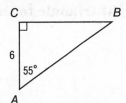

Find *m* ∠*J* for each right triangle to the nearest degree.

13.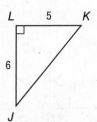

14.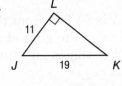

10-8 Practice

Trigonometric Ratios

Find the values of the three trigonometric ratios for angle A.

1.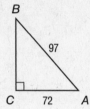

2.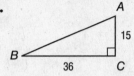

Use a calculator to find the value of each trigonometric ratio to the nearest ten-thousandth.

3. tan 26°

4. sin 53°

5. cos 81°

Solve each right triangle. Round each side length to the nearest tenth.

6.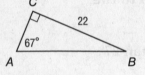

7.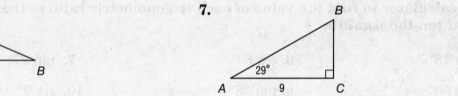

Find $m\angle J$ for each right triangle to the nearest degree.

8.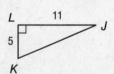

9.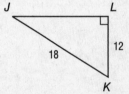

10. **SURVEYING** If point A is 54 feet from the tree, and the angle between the ground at point A and the top of the tree is 25°, find the height h of the tree.

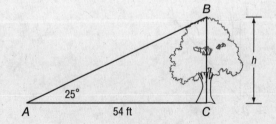

11-1 Skills Practice

Inverse Variation

Determine whether each table or equation represents an *inverse* or a *direct* variation. Explain.

1.

x	y
0.5	8
1	4
2	2
4	1

2. $xy = \dfrac{2}{3}$

3. $-2x + y = 0$

Assume that *y* varies inversely as *x*. Write an inverse variation equation that relates *x* and *y*. Then graph the equation.

4. $y = 2$ when $x = 5$

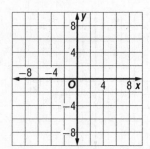

5. $y = -6$ when $x = -6$

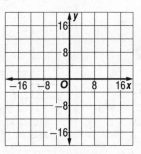

6. $y = -4$ when $x = -12$

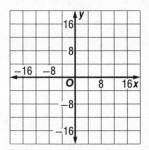

7. $y = 15$ when $x = 3$

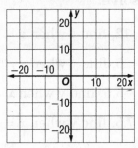

Solve. Assume that *y* varies inversely as *x*.

8. If $y = 4$ when $x = 8$,
find y when $x = 2$.

9. If $y = -7$ when $x = 3$,
find y when $x = -3$.

10. If $y = -6$ when $x = -2$,
find y when $x = 4$.

11. If $y = -24$ when $x = -3$,
find x when $y = -6$.

12. If $y = 15$ when $x = 1$,
find x when $y = -3$.

13. If $y = 48$ when $x = -4$,
find y when $x = 6$.

14. If $y = -4$ when $x = \dfrac{1}{2}$, find x when $y = 2$.

11-1 Practice

Inverse Variation

Determine whether each table or equation represents an *inverse* or a *direct* variation. Explain.

1.

x	y
0.25	40
0.5	20
2	5
8	1.25

2.

x	y
−2	8
0	0
2	−8
4	−16

3. $\frac{y}{x} = -3$

4. $y = \frac{7}{x}$

Asssume that *y* varies inversely as *x*. Write an inverse variation equation that relates *x* and *y*. Then graph the equation.

5. $y = -2$ when $x = -12$ **6.** $y = -6$ when $x = -5$ **7.** $y = 2.5$ when $x = 2$

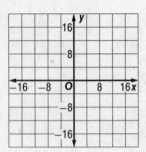
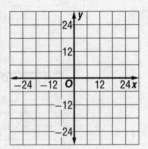
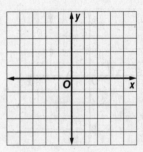

Write an inverse variation equation that relates *x* and *y*. Assume that *y* varies inversely as *x*. Then solve.

8. If $y = 124$ when $x = 12$, find y when $x = -24$.

9. If $y = -8.5$ when $x = 6$, find y when $x = -2.5$.

10. If $y = 3.2$ when $x = -5.5$, find y when $x = 6.4$.

11. If $y = 0.6$ when $x = 7.5$, find y when $x = -1.25$.

12. EMPLOYMENT The manager of a lumber store schedules 6 employees to take inventory in an 8-hour work period. The manager assumes all employees work at the same rate.

   a. Suppose 2 employees call in sick. How many hours will 4 employees need to take inventory?

   b. If the district supervisor calls in and says she needs the inventory finished in 6 hours, how many employees should the manager assign to take inventory?

13. TRAVEL Jesse and Joaquin can drive to their grandparents' home in 3 hours if they average 50 miles per hour. Since the road between the homes is winding and mountainous, their parents prefer they average between 40 and 45 miles per hour. How long will it take to drive to the grandparents' home at the reduced speed?

11-2 Skills Practice

Rational Functions

State the excluded value for each function.

1. $y = \dfrac{6}{x}$

2. $y = \dfrac{2}{x - 2}$

3. $y = \dfrac{x}{x + 6}$

4. $y = \dfrac{x - 3}{x + 4}$

5. $y = \dfrac{3x - 5}{x + 8}$

6. $y = \dfrac{-5}{2x - 14}$

7. $y = \dfrac{x}{3x + 21}$

8. $y = \dfrac{x - 1}{9x - 36}$

9. $y = \dfrac{9}{5x + 40}$

Identify the asymptotes of each function. Then graph the function.

10. $y = \dfrac{1}{x}$

11. $y = \dfrac{3}{x}$

12. $y = \dfrac{2}{x + 1}$

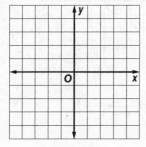
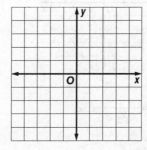
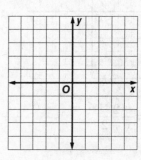

13. $y = \dfrac{3}{x - 2}$

14. $y = \dfrac{2}{x + 1} - 1$

15. $y = \dfrac{1}{x - 2} + 3$

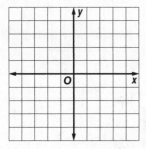
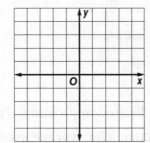
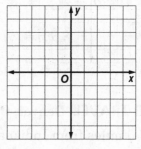

11-2 **Practice**

Rational Functions

State the excluded value for each function.

1. $y = \dfrac{-1}{x}$

2. $y = \dfrac{3}{x + 5}$

3. $y = \dfrac{2x}{x - 5}$

4. $y = \dfrac{x - 1}{12x + 36}$

5. $y = \dfrac{x + 1}{2x + 3}$

6. $y = \dfrac{1}{5x - 2}$

Identify the asymptotes of each function. Then graph the function.

7. $y = \dfrac{1}{x}$

8. $y = \dfrac{3}{x}$

9. $y = \dfrac{2}{x - 1}$

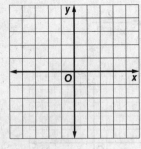
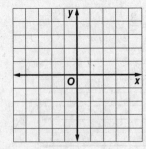
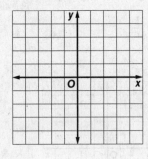

10. $y = \dfrac{2}{x + 2}$

11. $y = \dfrac{1}{x - 3} + 2$

12. $y = \dfrac{2}{x + 1} - 1$

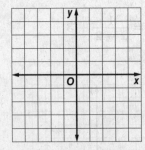
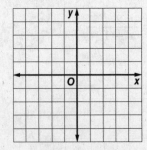
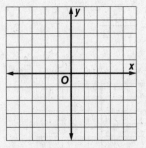

13. AIR TRAVEL Denver, Colorado, is located approximately 1000 miles from Indianapolis, Indiana. The average speed of a plane traveling between the two cities is given by $y = \dfrac{1000}{x}$, where x is the total flight time. Graph the function.

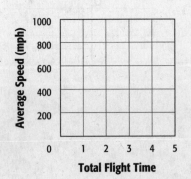

11-3 Skills Practice

Simplifying Rational Expressions

State the excluded values for each rational expression.

1. $\dfrac{2p}{p-7}$

2. $\dfrac{4n+1}{n+4}$

3. $\dfrac{k+2}{k^2-4}$

4. $\dfrac{3x+15}{x^2-25}$

5. $\dfrac{y^2-9}{y^2+3y-18}$

6. $\dfrac{b^2-2b-8}{b^2+7b+10}$

Simplify each expression. State the excluded values of the variables.

7. $\dfrac{21bc}{28bc^2}$

8. $\dfrac{12m^2r}{24mr^3}$

9. $\dfrac{16x^3y^2}{36x^5y^3}$

10. $\dfrac{8a^2b^3}{40a^3b}$

11. $\dfrac{n+6}{3n+18}$

12. $\dfrac{4x-4}{4x+4}$

13. $\dfrac{y^2-64}{y+8}$

14. $\dfrac{y^2-7y-18}{y-9}$

15. $\dfrac{z+1}{z^2-1}$

16. $\dfrac{x+6}{x^2+2x-24}$

17. $\dfrac{2d+10}{d^2-2d-35}$

18. $\dfrac{3h-9}{h^2-7h+12}$

19. $\dfrac{t^2+5t+6}{t^2+6t+8}$

20. $\dfrac{a^2+3a-4}{a^2+2a-8}$

21. $\dfrac{x^2+10x+24}{x^2-2x-24}$

22. $\dfrac{b^2-6b+9}{b^2-9b+18}$

11-3 Practice

Simplifying Rational Expressions

State the excluded values for each rational expression.

1. $\dfrac{4n - 28}{n^2 - 49}$

2. $\dfrac{p^2 - 16}{p^2 - 13p + 36}$

3. $\dfrac{a^2 - 2a - 15}{a^2 + 8a + 15}$

Simplify each expression. State the excluded values of the variables.

4. $\dfrac{12a}{48a^3}$

5. $\dfrac{6xyz^3}{3x^2y^2z}$

6. $\dfrac{36k^3np^2}{20k^2np^5}$

7. $\dfrac{5c^3d^4}{40cd^2 + 5c^4d^2}$

8. $\dfrac{p^2 - 8p + 12}{p - 2}$

9. $\dfrac{m^2 - 4m - 12}{m - 6}$

10. $\dfrac{m + 3}{m^2 - 9}$

11. $\dfrac{2b - 14}{b^2 - 9b + 14}$

12. $\dfrac{x^2 - 7x + 10}{x^2 - 2x - 15}$

13. $\dfrac{y^2 + 6y - 16}{y^2 - 4y + 4}$

14. $\dfrac{r^2 - 7r + 6}{r^2 + 6r - 7}$

15. $\dfrac{t^2 - 81}{t^2 - 12t + 27}$

16. $\dfrac{r^2 + r - 6}{r^2 + 4r - 12}$

17. $\dfrac{2x^2 + 18x + 36}{3x^2 - 3x - 36}$

18. $\dfrac{2y^2 + 9y + 4}{4y^2 - 4y - 3}$

19. **ENTERTAINMENT** Fairfield High spent d dollars for refreshments, decorations, and advertising for a dance. In addition, they hired a band for $550.

    a. Write an expression that represents the cost of the band as a fraction of the total amount spent for the school dance.

    b. If d is $1650, what percent of the budget did the band account for?

20. **PHYSICAL SCIENCE** Mr. Kaminksi plans to dislodge a tree stump in his yard by using a 6-foot bar as a lever. He places the bar so that 0.5 foot extends from the fulcrum to the end of the bar under the tree stump. In the diagram, b represents the total length of the bar and t represents the portion of the bar beyond the fulcrum.

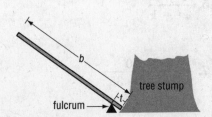

    a. Write an equation that can be used to calculate the mechanical advantage.

    b. What is the mechanical advantage?

    c. If a force of 200 pounds is applied to the end of the lever, what is the force placed on the tree stump?

11-4 Skills Practice

Multiplying and Dividing Rational Expressions

Find each product.

1. $\dfrac{14}{c^2} \cdot \dfrac{c^5}{2c}$

2. $\dfrac{3m^2}{2t} \cdot \dfrac{t^2}{12}$

3. $\dfrac{2a^2b}{b^2c} \cdot \dfrac{b}{a}$

4. $\dfrac{2x^2y}{3x^2y} \cdot \dfrac{3xy}{4y}$

5. $\dfrac{3(4m-6)}{18r} \cdot \dfrac{9r^2}{2(4m-6)}$

6. $\dfrac{4(n+2)}{n(n-2)} \cdot \dfrac{n-2}{n+2}$

7. $\dfrac{(y-3)(y+3)}{4} \cdot \dfrac{8}{y+3}$

8. $\dfrac{(x-2)(x+2)}{x(8x+3)} \cdot \dfrac{2(8x+3)}{x-2}$

9. $\dfrac{(a-7)(a+7)}{a(a+5)} \cdot \dfrac{a+5}{a+7}$

10. $\dfrac{4(b+4)}{(b-4)(b-3)} \cdot \dfrac{b-3}{b+4}$

Find each quotient.

11. $\dfrac{c^3}{d^3} \div \dfrac{d^3}{c^3}$

12. $\dfrac{x^3}{y^2} \div \dfrac{x^3}{y}$

13. $\dfrac{6a^3}{4f^2} \div \dfrac{2a^2}{12f^2}$

14. $\dfrac{4m^3}{rp^2} \div \dfrac{2m}{rp}$

15. $\dfrac{3b+3}{b+2} \div (b+1)$

16. $\dfrac{x-5}{x+3} \div (x-5)$

17. $\dfrac{x^2-x-12}{6} \div \dfrac{x+3}{x-4}$

18. $\dfrac{a^2-5a-6}{3} \div \dfrac{a-6}{a+1}$

19. $\dfrac{m^2+2m+1}{10m-10} \div \dfrac{m+1}{20}$

20. $\dfrac{y^2+10y+25}{3y-9} \div \dfrac{y+5}{y-3}$

21. $\dfrac{b+4}{b^2-8b+16} \div \dfrac{2b+8}{b-8}$

22. $\dfrac{6x+6}{x-1} \div \dfrac{x^2+3x+2}{2x-2}$

11-4 Practice

Multiplying and Dividing Rational Expressions

Find each product or quotient.

1. $\dfrac{18x^2}{10y^2} \cdot \dfrac{15y^3}{24x}$

2. $\dfrac{24rt^2}{8r^4t^3} \cdot \dfrac{12r^3t^2}{36r^2t}$

3. $\dfrac{(x+2)(x+2)}{8} \cdot \dfrac{72}{(x+2)(x-2)}$

4. $\dfrac{m+7}{(m-6)(m+2)} \cdot \dfrac{(m-6)(m+4)}{(m+7)}$

5. $\dfrac{a-4}{a^2-a-12} \cdot \dfrac{a+3}{a-6}$

6. $\dfrac{4x+8}{x^2} \cdot \dfrac{x}{x^2-5x-14}$

7. $\dfrac{n^2+10n+16}{5n-10} \cdot \dfrac{n-2}{n^2+9n+8}$

8. $\dfrac{3y-9}{y^2-9y+20} \cdot \dfrac{y^2-8y+16}{y-3}$

9. $\dfrac{b^2+5b+4}{b^2-36} \cdot \dfrac{b^2+5b-6}{b^2+2b-8}$

10. $\dfrac{t^2+6t+9}{t^2-10t+25} \cdot \dfrac{t^2-t-20}{t^2+7t+12}$

11. $\dfrac{28a^2}{7b^2} \div \dfrac{21a^3}{35b}$

12. $\dfrac{mn^2p^3}{x^4y^2} \div \dfrac{mnp^2}{x^3y}$

13. $\dfrac{2a}{a-1} \div (a+1)$

14. $\dfrac{z^2-16}{3z} \div (z-4)$

15. $\dfrac{4y+20}{y-3} \div \dfrac{y+5}{2y-6}$

16. $\dfrac{4x+12}{6x-24} \div \dfrac{2x+6}{x+3}$

17. $\dfrac{b^2+2b-8}{b^2-11b+18} \div \dfrac{2b-8}{2b-18}$

18. $\dfrac{3x-3}{x^2-6x+9} \div \dfrac{6x-6}{x^2-5x+6}$

19. $\dfrac{a^2+8a+12}{a^2-7a+10} \div \dfrac{a^2-4a-12}{a^2+3a-10}$

20. $\dfrac{y^2+6y-7}{y^2+8y-9} \div \dfrac{y^2+9y+14}{y^2+7y-18}$

21. **BIOLOGY** The heart of an average person pumps about 9000 liters of blood per day. How many quarts of blood does the heart pump per hour? (*Hint:* One quart is equal to 0.946 liter.) Round to the nearest whole number.

22. **TRAFFIC** On Saturday, it took Ms. Torres 24 minutes to drive 20 miles from her home to her office. During Friday's rush hour, it took 75 minutes to drive the same distance.

   a. What was Ms. Torres's average speed in miles per hour on Saturday?

   b. What was her average speed in miles per hour on Friday?

11-5 Skills Practice

Dividing Polynomials

Find each quotient.

1. $(20x^2 + 12x) \div 4x$

2. $(18n^2 + 6n) \div 3n$

3. $(b^2 - 12b + 5) \div 2b$

4. $(8r^2 + 5r - 20) \div 4r$

5. $\dfrac{12p^3r^2 + 18p^2r - 6pr}{6p^2r}$

6. $\dfrac{15k^2u - 10ku + 25u^2}{5ku}$

7. $(x^2 - 5x - 6) \div (x - 6)$

8. $(a^2 - 10a + 16) \div (a - 2)$

9. $(n^2 - n - 20) \div (n + 4)$

10. $(y^2 + 4y - 21) \div (y - 3)$

11. $(h^2 - 6h + 9) \div (h - 2)$

12. $(b^2 + 5b - 2) \div (b + 6)$

13. $(y^2 + 6y + 1) \div (y + 2)$

14. $(m^2 - 2m - 5) \div (m - 3)$

15. $\dfrac{2c^2 - 5c - 3}{2c + 1}$

16. $\dfrac{2r^2 + 6r - 20}{2r - 4}$

17. $\dfrac{x^3 - 3x^2 - 6x - 20}{x - 5}$

18. $\dfrac{p^3 - 4p^2 + p + 6}{p - 2}$

19. $\dfrac{n^3 - 6n - 2}{n + 1}$

20. $\dfrac{y^3 - y^2 - 40}{y - 4}$

11-5 Practice

Dividing Polynomials

Find each quotient.

1. $(6q^2 - 18q - 9) \div 9q$

2. $(y^2 + 6y + 2) \div 3y$

3. $\dfrac{12a^2b - 3ab^2 + 42ab}{6a^2b}$

4. $\dfrac{2m^3p^2 + 56mp - 4m^2p^3}{8m^3p}$

5. $(x^2 - 3x - 40) \div (x + 5)$

6. $(3m^2 - 20m + 12) \div (m - 6)$

7. $(a^2 + 5a + 20) \div (a - 3)$

8. $(x^2 - 3x - 2) \div (x + 7)$

9. $(t^2 + 9t + 28) \div (t + 3)$

10. $(n^2 - 9n + 25) \div (n - 4)$

11. $\dfrac{6r^2 - 5r - 56}{3r + 8}$

12. $\dfrac{20w^2 + 39w + 18}{5w + 6}$

13. $(x^3 + 2x^2 - 16) \div (x - 2)$

14. $(t^3 - 11t - 6) \div (t + 3)$

15. $\dfrac{x^3 + 6x^2 + 3x + 1}{x - 2}$

16. $\dfrac{6d^3 + d^2 - 2d + 17}{2d + 3}$

17. $\dfrac{2k^3 + k^2 - 12k + 11}{2k - 3}$

18. $\dfrac{9y^3 - y - 1}{3y + 2}$

19. LANDSCAPING Jocelyn is designing a bed for cactus specimens at a botanical garden. The total area can be modeled by the expression $2x^2 + 7x + 3$, where x is in feet.

  a. Suppose in one design the length of the cactus bed is $4x$, and in another, the length is $2x + 1$. What are the widths of the two designs?

  b. If $x = 3$ feet, what will be the dimensions of the cactus bed in each of the designs?

20. FURNITURE Teri is upholstering the seats of four chairs and a bench. She needs $\frac{1}{4}$ square yard of fabric for each chair, and $\frac{1}{2}$ square yard for the bench. If the fabric at the store is 45 inches wide, how many yards of fabric will Teri need to cover the chairs and the bench if there is no waste?

11-6 Skills Practice

Adding and Subtracting Rational Expressions

Find each sum or difference.

1. $\dfrac{2y}{5} + \dfrac{y}{5}$

2. $\dfrac{4r}{9} + \dfrac{5r}{9}$

3. $\dfrac{t+3}{7} - \dfrac{t}{7}$

4. $\dfrac{c+8}{4} - \dfrac{c+6}{4}$

5. $\dfrac{x+2}{3} + \dfrac{x+5}{3}$

6. $\dfrac{g+2}{4} + \dfrac{g-8}{4}$

7. $\dfrac{x}{x-1} - \dfrac{1}{x-1}$

8. $\dfrac{3r}{r+3} - \dfrac{r}{r+3}$

Find the LCM of each pair of polynomials.

9. $4x^2y,\ 12xy^2$

10. $n+2,\ n-3$

11. $2r-1,\ r+4$

12. $t+4,\ 4t+16$

Find each sum or difference.

13. $\dfrac{5}{4r} - \dfrac{2}{r^2}$

14. $\dfrac{5x}{3y^2} - \dfrac{2x}{9y}$

15. $\dfrac{x}{x+2} - \dfrac{4}{x-1}$

16. $\dfrac{d-1}{d-2} - \dfrac{3}{d+5}$

17. $\dfrac{b}{b-1} + \dfrac{2}{b-4}$

18. $\dfrac{k}{k-5} + \dfrac{k-1}{k+5}$

19. $\dfrac{3x+15}{x^2-25} + \dfrac{x}{x+5}$

20. $\dfrac{x-3}{x^2-4x+4} + \dfrac{x+2}{x-2}$

11-6 Practice

Adding and Subtracting Rational Expressions

Find each sum or difference.

1. $\dfrac{n}{8} + \dfrac{3n}{8}$

2. $\dfrac{7u}{16} + \dfrac{5u}{16}$

3. $\dfrac{w+9}{9} + \dfrac{w+4}{9}$

4. $\dfrac{x-6}{2} - \dfrac{x-7}{2}$

5. $\dfrac{n+14}{5} - \dfrac{n-14}{5}$

6. $\dfrac{6}{c-1} - \dfrac{-2}{c-1}$

7. $\dfrac{x-5}{x+2} + \dfrac{-2}{x+2}$

8. $\dfrac{r+5}{r-5} + \dfrac{2r-1}{r-5}$

9. $\dfrac{4p+14}{p+4} + \dfrac{2p+10}{p+4}$

Find the LCM of each pair of polynomials.

10. $3a^3b^2,\ 18ab^3$

11. $w-4,\ w+2$

12. $5d-20,\ d-4$

13. $6p+1,\ p-1$

14. $x^2+5x+4,\ (x+1)^2$

15. $m^2+3m-10,\ m^2-4$

Find each sum or difference.

16. $\dfrac{6p}{5x^2} - \dfrac{2p}{3x}$

17. $\dfrac{m+4}{m-3} - \dfrac{2}{m-6}$

18. $\dfrac{y+3}{y^2-16} + \dfrac{3y-2}{y^2+8y+16}$

19. $\dfrac{p+1}{p^2+3p-4} + \dfrac{p}{p+4}$

20. $\dfrac{t+3}{t^2-3t-10} - \dfrac{4t-8}{t^2-10t+25}$

21. $\dfrac{4y}{y^2-y-6} - \dfrac{3y+3}{y^2-4}$

22. SERVICE Members of the ninth grade class at Pine Ridge High School are organizing into service groups. What is the minimum number of students who must participate for all students to be divided into groups of 4, 6, or 9 students with no one left out?

23. GEOMETRY Find an expression for the perimeter of rectangle $ABCD$. Use the formula $P = 2\ell + 2w$.

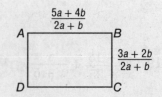

11-7 Skills Practice

Mixed Expressions and Complex Fractions

Write each mixed expression as a rational expression.

1. $6 + \dfrac{4}{h}$

2. $7 + \dfrac{6}{p}$

3. $4b + \dfrac{b}{c}$

4. $8q - \dfrac{2q}{r}$

5. $2 + \dfrac{4}{d-5}$

6. $5 - \dfrac{6}{f+2}$

7. $b^2 + \dfrac{12}{b+3}$

8. $m - \dfrac{6}{m-7}$

9. $2a + \dfrac{a-2}{a}$

10. $4r - \dfrac{r+9}{2r}$

Simplify each expression.

11. $\dfrac{2\frac{1}{2}}{4\frac{3}{4}}$

12. $\dfrac{3\frac{2}{3}}{5\frac{2}{5}}$

13. $\dfrac{\frac{r}{n^2}}{\frac{r^2}{n}}$

14. $\dfrac{\frac{a^2}{b^3}}{\frac{a}{b}}$

15. $\dfrac{\frac{x^2y}{c}}{\frac{xy^3}{c^2}}$

16. $\dfrac{\frac{r-2}{r+3}}{\frac{r-2}{3}}$

17. $\dfrac{\frac{w+4}{w}}{\frac{w^2-16}{w}}$

18. $\dfrac{\frac{x^2-1}{x}}{\frac{x-1}{x^2}}$

19. $\dfrac{\frac{b^2-4}{b^2+7b+10}}{b-2}$

20. $\dfrac{\frac{k^2+5k+6}{k^2-9}}{k+2}$

21. $\dfrac{g+\frac{12}{g+8}}{g+6}$

22. $\dfrac{p+\frac{9}{p-6}}{p-3}$

11-7 Practice

Mixed Expressions and Complex Fractions

Write each mixed expression as a rational expression.

1. $14 - \dfrac{9}{u}$

2. $7d + \dfrac{4d}{c}$

3. $3n + \dfrac{6 - n}{n}$

4. $5b - \dfrac{b + 3}{2b}$

5. $3 + \dfrac{t + 5}{t^2 - 1}$

6. $2a + \dfrac{a - 1}{a + 1}$

7. $2p + \dfrac{p + 1}{p - 3}$

8. $4n^2 + \dfrac{n - 1}{n^2 - 1}$

9. $(t + 1) + \dfrac{4}{t + 5}$

Simplify each expression.

10. $\dfrac{3\frac{2}{5}}{2\frac{5}{6}}$

11. $\dfrac{\frac{m^2}{6p}}{\frac{3m}{p^2}}$

12. $\dfrac{\frac{x^2 - y^2}{x^2}}{\frac{x + y}{3x}}$

13. $\dfrac{\frac{a - 4}{a^2}}{\frac{a^2 - 16}{a}}$

14. $\dfrac{\frac{q^2 - 7q + 12}{q^2 - 16}}{q - 3}$

15. $\dfrac{\frac{k^2 + 6k}{k^2 + 4k - 5}}{\frac{k - 8}{k^2 - 9k + 8}}$

16. $\dfrac{\frac{b^2 + b - 12}{b^2 + 3b - 4}}{\frac{b - 3}{b^2 - b}}$

17. $\dfrac{g - \frac{10}{g + 9}}{g - \frac{5}{g + 4}}$

18. $\dfrac{y + \frac{6}{y - 7}}{y - \frac{7}{y + 6}}$

19. TRAVEL Ray and Jan are on a $12\frac{1}{2}$-hour drive from Springfield, Missouri, to Chicago, Illinois. They stop for a break every $3\frac{1}{4}$ hours.

    a. Write an expression to model this situation.

    b. How many stops will Ray and Jan make before arriving in Chicago?

20. CARPENTRY Tai needs several $2\frac{1}{4}$-inch wooden rods to reinforce the frame on a futon. She can cut the rods from a $24\frac{1}{2}$-inch dowel purchased from a hardware store. How many wooden rods can she cut from the dowel?

11-8 Skills Practice

Rational Equations

Solve each equation. State any extraneous solutions.

1. $\dfrac{5}{c} = \dfrac{2}{c+3}$

2. $\dfrac{3}{q} = \dfrac{5}{q+4}$

3. $\dfrac{7}{m+1} = \dfrac{12}{m+2}$

4. $\dfrac{3}{x+2} = \dfrac{5}{x+8}$

5. $\dfrac{y}{y-2} = \dfrac{y+1}{y-5}$

6. $\dfrac{b-2}{b} = \dfrac{b+4}{b+2}$

7. $\dfrac{3m}{2} - \dfrac{1}{4} = \dfrac{10m}{8}$

8. $\dfrac{7g}{9} + \dfrac{1}{3} = \dfrac{5g}{6}$

9. $\dfrac{2a+5}{6} - \dfrac{2a}{3} = -\dfrac{1}{2}$

10. $\dfrac{n-3}{10} + \dfrac{n-5}{5} = \dfrac{1}{2}$

11. $\dfrac{c+2}{c} + \dfrac{c+3}{c} = 7$

12. $\dfrac{3b-4}{b} - \dfrac{b-7}{b} = 1$

13. $\dfrac{m-4}{m} - \dfrac{m-11}{m+4} = \dfrac{1}{m}$

14. $\dfrac{f+2}{f} - \dfrac{f+1}{f+5} = \dfrac{1}{f}$

15. $\dfrac{r+3}{r-1} - \dfrac{r}{r-3} = 0$

16. $\dfrac{u+1}{u-2} - \dfrac{u}{u+1} = 0$

17. $\dfrac{-2}{x+1} + \dfrac{2}{x} = 1$

18. $\dfrac{5}{m-4} - \dfrac{m}{2m-8} = 1$

19. ACTIVISM Maury and Tyra are making phone calls to state representatives' offices to lobby for an issue. Maury can call all 120 state representatives in 10 hours. Tyra can call all 120 state representatives in 8 hours. How long would it take them to call all 120 state representatives together?

11-8 Practice

Rational Equations

Solve each equation. State any extraneous solutions.

1. $\dfrac{5}{n+2} = \dfrac{7}{n+6}$

2. $\dfrac{x}{x-5} = \dfrac{x+4}{x-6}$

3. $\dfrac{k+5}{k} = \dfrac{k-1}{k+9}$

4. $\dfrac{2h}{h-1} = \dfrac{2h+1}{h+2}$

5. $\dfrac{4y}{3} + \dfrac{1}{2} = \dfrac{5y}{6}$

6. $\dfrac{y-2}{4} - \dfrac{y+2}{5} = -1$

7. $\dfrac{2q-1}{6} - \dfrac{q}{3} = \dfrac{q+4}{18}$

8. $\dfrac{5}{p-1} - \dfrac{3}{p+2} = 0$

9. $\dfrac{3t}{3t-3} - \dfrac{1}{9t+3} = 1$

10. $\dfrac{4x}{2x+1} - \dfrac{2x}{2x+3} = 1$

11. $\dfrac{d-3}{d} - \dfrac{d-4}{d-2} = \dfrac{1}{d}$

12. $\dfrac{3y-2}{y-2} + \dfrac{y^2}{2-y} = -3$

13. $\dfrac{2}{m+2} - \dfrac{m+2}{m-2} = \dfrac{7}{3}$

14. $\dfrac{n+2}{n} + \dfrac{n+5}{n+3} = -\dfrac{1}{n}$

15. $\dfrac{1}{z+1} - \dfrac{6-z}{6z} = 0$

16. $\dfrac{2p}{p-2} + \dfrac{p+2}{p^2-4} = 1$

17. $\dfrac{x+7}{x^2-9} - \dfrac{x}{x+3} = 1$

18. $\dfrac{2n}{n-4} - \dfrac{n+6}{n^2-16} = 1$

19. **PUBLISHING** Tracey and Alan publish a 10-page independent newspaper once a month. At production, Alan usually spends 6 hours on the layout of the paper. When Tracey helps, layout takes 3 hours and 20 minutes.

   a. Write an equation that could be used to determine how long it would take Tracey to do the layout by herself.

   b. How long would it take Tracey to do the job alone?

20. **TRAVEL** Emilio made arrangements to have Lynda pick him up from an auto repair shop after he dropped his car off. He called Lynda to tell her he would start walking and to look for him on the way. Emilio and Lynda live 10 miles from the auto shop. It takes Emilio $2\frac{1}{4}$ hours to walk the distance and Lynda 15 minutes to drive the distance.

   a. If Emilio and Lynda leave at the same time, when should Lynda expect to spot Emilio on the road?

   b. How far will Emilio have walked when Lynda picks him up?

12-1 Skills Practice

Designing a Survey

Identify each sample, and suggest a population from which it was selected. Then classify the type of data collection used.

1. **LANDSCAPING** A homeowner is concerned about the quality of the topsoil in her back yard. The back yard is divided into 5 equal sections, and then a 1-inch plug of topsoil is randomly removed from each of the 5 sections. The soil is taken to a nursery and analyzed for mineral content.

2. **HEALTH** A hospital's administration is interested in opening a gym on the premises for all its employees. They ask each member of the night-shift emergency room staff if he or she would use the gym, and if so, what hours the employee would prefer to use it.

3. **POLITICS** A senator wants to know her approval rating among the constituents in her state. She sends questionnaires to the households of 1000 registered voters.

Identify each sample as *biased* or *unbiased*. Explain your reasoning.

4. **MANUFACTURING** A company that produces motherboards for computers randomly selects 25 boxed motherboards out of a shipment of 1500, and then tests each selected motherboard to see that it meets specifications.

5. **GOVERNMENT** The first 100 people entering a county park on Thursday are asked their opinions on a proposed county ordinance that would allow dogs in county parks to go unleashed in certain designated areas.

Identify the sample and suggest a population from which it was selected. Then classify the sample as *simple*, *stratified*, or *systematic*. Explain your reasoning.

6. **MUSIC** To determine the music preferences of their customers, the owners of a music store randomly choose 10 customers to participate in an in-store interview in which they listen to new CDs from artists in all music categories.

7. **LIBRARIES** A community library asks every tenth patron who enters the library to name the type or genre of book he or she is most likely to borrow. They conduct the interviews from opening to closing on three days of the week. They will use the data for new acquisitions.

8. **COMPUTERS** To determine the number of students who use computers at home, the high school office chooses 10 students at random from each grade, and then interviews the students.

12-1 Practice

Designing a Survey

Identify each sample, and suggest a population from which it was selected. Then classify the type of data collection used.

1. **GOVERNMENT** At a town council meeting, the chair asks 5 citizens attending for their opinions on whether to approve rezoning for a residential area.

2. **BOTANY** To determine the extent of leaf blight in the maple trees at a nature preserve, a botanist divides the reserve into 10 sections, randomly selects a 200-foot by 200-foot square in the section, and then examines all the maple trees in the section.

3. **FINANCES** To determine the popularity of online banking in the United States, a polling company sends a mail-in survey to 5000 adults to see if they bank online, and if they do, how many times they bank online each month.

Identify each sample as *biased* or *unbiased*. Explain your reasoning.

4. **SHOES** A shoe manufacturer wants to check the quality of its shoes. Every twenty minutes, 20 pairs of shoes are pulled off the assembly line for a quality inspection.

5. **BUSINESS** To learn which benefits employees at a large company think are most important, the management has a computer select 50 employees at random. The employees are then interviewed by the Human Relations department.

For Question 6, identify the sample, and suggest a population from which it was selected. Then classify the sample as *simple*, *stratified*, or *systematic*. Explain your reasoning.

6. **BUSINESS** An insurance company checks every hundredth claim payment to ensure that claims have been processed correctly.

7. **ENVIRONMENT** Suppose you want to know if a manufacturing plant is discharging contaminants into a local river. Describe an unbiased way in which you could check the river water for contaminants.

8. **SCHOOL** Suppose you want to know the issues most important to teachers at your school. Describe an unbiased way in which you could conduct your survey.

12-2 Skills Practice

Analyzing Survey Results

Which measure of central tendency best represents the data? Justify your answer. Then find the measure.

1. **SNOWFALL** A weather station keeps records of how many inches of snow fall each week: {9, 2, 0, 3, 0, 2, 1, 2, 3, 1}.

2. **SALES** A supermarket keeps records of how many boxes of cereal are sold each day in a week: {12, 9, 11, 14, 19, 49, 18}.

3. **ELECTIONS** A city councilman keeps track of the number of votes he receives in each district: {68, 66, 58, 59, 61, 62, 67}.

Given the following portion of a survey report, evaluate the validity of the information and conclusion.

4. **ECONOMY** The Gallup polling company interviewed 1464 U.S. adults nationwide.
Question: How would you rate economic conditions in this country today?
Results: excellent, 3%; good, 22%; only fair, 44%; poor 32%
Conclusion: Americans have confidence in the economy.

5. **DOGS** A pet store surveyed its customers to find their favorite breed of dog.
Question: What is your favorite breed of dog?
Results: golden retriever, 26%; collie, 19%; terrier, 11%; bulldog, 8%; pug, 24%; other, 12%
Conclusion: The golden retriever is the favorite dog of most customers.

Determine whether each display gives an accurate picture of the survey results.

6. **TRASH INCINERATORS** A local newspaper surveyed 530 randomly chosen Eastwich residents.
Question: Do you support closing the trash incinerator in Eastwich?
Conclusion: Eastwich residents overwhelmingly support closing the trash incinerator.

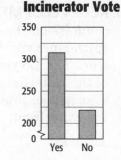

Incinerator Vote

7. **ISSUES** A television station interviewed 400 randomly chosen voters.
Question: What issue matters most to you in choosing a candidate to vote for?
Conclusion: Most voters do not care about the environment.

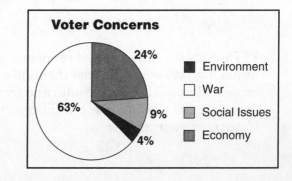

Voter Concerns
24% — Environment
63% — War
9% — Social Issues
4% — Economy

12-2 Practice

Analyzing Survey Results

Which measure of central tendency best represents the data? Justify your answer. Then find the measure.

1. **CALCULATORS** The math department counts how many graphing calculators are in each classroom: {20, 19, 20, 20, 18, 19, 20, 18, 19}.

2. **BUDGETING** The Brady family keeps track of its monthly electric bills: {$134, $122, $128, $127, $136, $120, $129}.

3. **AUTOMATED TELLERS** A bank keeps track of how many customers use its ATM each hour: {39, 42, 44, 120, 54, 48, 43}.

Given the following portion of a survey report, evaluate the validity of the information and conclusion.

4. **HOMEWORK** Chris polled 16 of his friends during study hall.
   Question: Do teachers at Edison High School assign too much homework?
   Results: yes, 94%; no, 6%
   Conclusion: Teachers at Edison High School should assign less homework.

5. **SMOKING** SurveyUSA polled 500 randomly selected adults in Kentucky.
   Question: Do you want to see smoking banned from restaurants, bars, and most indoor public places in Kentucky?
   Results: banned, 58%; allowed, 41%; not sure, 1%
   Conclusion: The United States should ban smoking indoors.

Determine whether the display gives an accurate picture of the survey results.

6. **REDEVELOPMENT** A local news broadcast commissioned a poll of 600 randomly chosen Providence residents.
   Question: Do you support or oppose the redevelopment of the waterfront?
   Conclusion: Providence residents support redeveloping the waterfront.

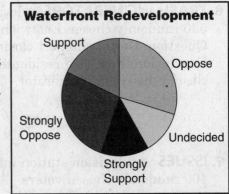

7. **PETS** Ernesto took a poll of randomly selected students at his high school and asked them how many pets they owned. He recorded the results and made the graph shown at the right. Write a valid conclusion using data to support your answer.

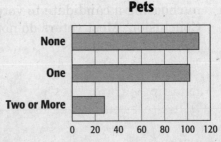

12-3 / Skills Practice

Statistics and Parameters

Identify the sample and the population for each situation. Then describe the sample statistic and the population parameter.

1. **RESTAURANTS** A restaurant randomly selects 10 patrons on Saturday night. The median amount spent on beverages is then calculated for the sample.

2. **KITTENS** A veterinarian randomly selects 3 kittens from a litter. The mean weight of the 3 kittens is calculated.

3. **PRODUCE** A produce clerk randomly selects 20 bags of apples from each week's shipment and counts the total number of apples in each bag. The mode number of apples is calculated for the sample.

Find the mean absolute deviation.

4. **WILDLIFE** A researcher counts the number of river otters observed on each acre of land in a state park: {0, 10, 14, 6, 0, 8, 4}.

5. **FISHING** A fisherman records the weight of each black bass he catches during a fishing trip: {12, 7, 8, 13, 6, 14}.

6. **BUDGETING** Xavier keeps track of how much money he spends on gasoline each week: {20, 13, 26, 0, 33, 16, 18}.

Find the mean, variance, and standard deviation of each set of data.

7. {2, 0, 10, 4}

8. {6, 7, 6, 9}

9. {10, 9, 13, 6, 7}

10. {6, 8, 2, 3, 2, 9}

11. {23, 18, 28, 26, 15}

12. {44, 35, 50, 37, 43, 38, 40}

13. **PARKING** A city councilor wants to know how much revenue the city would earn by installing parking meters on Main Street. He counts the number of cars parked on Main Street each weekday: {64, 79, 81, 53, 63}. Find the standard deviation.

12-3 Practice

Statistics and Parameters

Identify the sample and the population for each situation. Then describe the sample statistic and the population parameter.

1. MARINE BIOLOGY A marine biologist randomly selects 30 oysters from a research tank. The mean weight of the 30 oysters is calculated.

2. CIVIL ENGINEERING A civic engineer randomly selects 5 city intersections with traffic lights. The median length of a red light is calculated for the sample.

3. BASEBALL A baseball commissioner randomly selects 10 home games played by a major league team. The median attendance is calculated for the games in the sample.

Find the mean absolute deviation.

4. INVESTING A stock broker keeps a record of the daily closing price of a share of stock in Bicsomm Corporation: {45.20, 46.10, 46.85, 42.55, 40.80}.

5. GOLF A golfer keeps track of his scores for each round: {78, 81, 86, 77, 75}.

6. WEATHER A meteorologist keeps track of the number of thunderstorms occuring each month in Sussex County: {0, 4, 7, 1, 3, 5, 2}.

Find the mean, variance, and standard deviation of each set of data.

7. {6, 11, 16, 9}

8. {2, 5, 8, 11, 4}

9. {23.4, 16.8, 9.7, 22.1}

10. $\{1, \frac{5}{2}, 4, \frac{11}{2}, \frac{1}{2}, 3\}$

11. {145, 166, 171, 150, 88}

12. {13, 24, 22, 17, 14, 29, 15, 22}

13. QUALITY CONTROL An inspector checks each automobile that comes off of the assembly line. He keeps a record of the number of defective cars each day: {3, 1, 2, 0, 0, 4, 3, 6, 1, 2}. Find the standard deviation.

12-4 Skills Practice

Permutations and Combinations

Use the Fundamental Counting Principle to evaluate each of the following.

1. SCHOOL PLAY Joseph and eight friends are attending the school play. How many ways can Joseph and his friends sit in 9 empty seats?

2. VIDEOS Sanjay is arranging his 6 favorite videos on a shelf. In how many ways can he do this?

Evaluate each expression.

3. $P(5, 2)$ **4.** $P(6, 4)$ **5.** $P(7, 3)$

6. $P(9, 4)$ **7.** $P(7, 5)$ **8.** $P(5, 3)$

9. $C(6, 2)$ **10.** $C(9, 7)$ **11.** $C(8, 4)$

12. $C(7, 5)$ **13.** $C(12, 2)$ **14.** $C(13, 7)$

15. $C(11, 2)$ **16.** $P(5, 4)$ **17.** $C(14, 5)$

18. $C(11, 6)$ **19.** $P(4, 2)$ **20.** $C(8, 6)$

12-4 Practice

Permutations and Combinations

Use the Fundamental Counting Principle to evaluate each of the following.

1. **ERRANDS** Wesley needs to stop at 6 stores on the way home from work. How many ways can Wesley arrange the 6 stops he needs to make?

2. **VOTING** There are 8 people waiting in line to cast their votes. How many ways can the people line up to vote?

Evaluate each expression.

3. $P(11, 3)$ 4. $P(6, 3)$ 5. $P(15, 3)$

6. $C(10, 9)$ 7. $C(12, 9)$ 8. $C(7, 3)$

9. $C(7, 4)$ 10. $C(12, 4)$ 11. $P(13, 3)$

12. $C(16, 12)$ 13. $C(17, 2)$ 14. $C(16, 15)$

15. $P(20, 5)$ 16. $P(11, 7)$ 17. $P(13, 1)$

18. $C(19, 16)$ 19. $P(15, 4)$ 20. $C(14, 7)$

21. **SPORTS** In how many orders can the top five finishers in a race finish?

22. **JUDICIAL PROCEDURE** The court system in a community needs to assign 3 out of 8 judges to a docket of criminal cases. Five of the judges are male and three are female.

    a. Does the selection of judges involve a permutation or a combination?

    b. In how many ways could three judges be chosen?

    c. If the judges are chosen randomly, what is the probability that all 3 judges are male?

12-5 Skills Practice

Probability of Compound Events

A bag contains 2 green, 9 brown, 7 yellow, and 4 blue marbles. Once a marble is selected, it is not replaced. Find each probability.

1. P(brown, then yellow)

2. P(green, then blue)

3. P(yellow, then yellow)

4. P(blue, then blue)

5. P(green, then *not* blue)

6. P(brown, then *not* green)

A die is rolled and a spinner like the one at the right is spun. Find each probability.

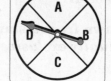

7. P(4 and A)

8. P(an even number and C)

9. P(2 or 5 and B or D)

10. P(a number less than 5 and B, C, or D)

A card is being drawn from a standard deck of playing cards. Determine whether the events are *mutually exclusive* or *not* mutually exclusive. Then find the probability.

11. P(jack or ten)

12. P(red or black)

13. P(queen or club)

14. P(red or ace)

15. P(diamond or black)

16. P(face card or spade)

Tiles numbered 1 through 20 are placed in a box. Tiles numbered 11 through 30 are placed in a second box. The first tile is randomly drawn from the first box. The second tile is randomly drawn from the second box. Find each probability.

17. P(both are greater than 15)

18. The first tile is odd and the second tile is less than 25.

19. The first tile is a multiple of 6 and the second tile is a multiple of 4.

20. The first tile is less than 15 and the second tile is even or greater than 25.

12-5 Practice

Probability of Compound Events

A bag contains 5 red, 3 brown, 6 yellow, and 2 blue marbles. Once a marble is selected, it is not replaced. Find each probability.

1. P(brown, then yellow, then red)

2. P(red, then red, then blue)

3. P(yellow, then yellow, then *not* blue)

4. P(brown, then brown, then *not* yellow)

A die is rolled and a card is drawn from a standard deck of 52 cards. Find each probability.

5. P(6 and king)

6. P(odd number and black)

7. P(less than 3 and heart)

8. P(greater than 1 and black ace)

A card is being drawn from a standard deck of playing cards. Determine whether the events are *mutually exclusive* or *not* mutually exclusive. Then find the probability.

9. P(spade or numbered card)

10. P(ace or red queen)

11. P(red or *not* face card)

12. P(heart or *not* queen)

Tiles numbered 1 through 25 are placed in a box. Tiles numbered 11 through 30 are placed in a second box. The first tile is randomly drawn from the first box. The second tile is randomly drawn from the second box. Find each probability.

13. P(both are greater than 15 and less than 20)

14. The first tile is greater than 10 and the second tile is less than 25 or even.

15. The first tile is a multiple of 3 or prime and the second tile is a multiple of 5.

16. The first tile is less than 9 or odd and the second tile is a multiple of 4 or less than 21.

17. WEATHER The forecast predicts a 40% chance of rain on Tuesday and a 60% chance on Wednesday. If these probabilities are independent, what is the chance that it will rain on both days?

18. FOOD Tomaso places favorite recipes in a bag for 4 pasta dishes, 5 casseroles, 3 types of chili, and 8 desserts.

    a. If Tomaso chooses one recipe at random, what is the probability that he selects a pasta dish or a casserole?

    b. If Tomaso chooses one recipe at random, what is the probability that he does *not* select a dessert?

    c. If Tomaso chooses two recipes at random without replacement, what is the probability that the first recipe he selects is a casserole and the second recipe he selects is a dessert?

12-6 Skills Practice

Probability Distributions

For Exercises 1–3, the spinner shown is spun three times.

1. Write the sample space with all possible outcomes.

2. Find the probability distribution X, where X represents the number of times the spinner lands on green for $X = 0$, $X = 1$, $X = 2$, and $X = 3$.

3. Make a probability graph of the data.

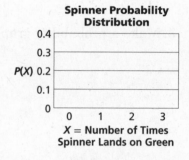

Spinner Probability Distribution

For Exercises 4–6, the spinner shown is spun two times.

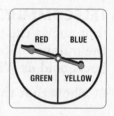

4. Write the sample space with all possible outcomes.

5. Find the probability distribution X, where X represents the number of times the spinner lands on yellow for $X = 0$, $X = 1$, and $X = 2$.

6. Make a probability graph of the data.

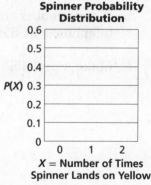

Spinner Probability Distribution

7. **BUSINESS** Use the table that shows the probability distribution of the number of minutes a customer spends at the express checkout at a supermarket.

X = Minutes	1	2	3	4	5 +
Probability	0.09	0.13	0.28	0.32	0.18

a. Show that the distribution is valid.

b. What is the probability that a customer spends less than 3 minutes at the checkout?

c. What is the probability that the customer spends at least 4 minutes at the checkout?

12-6 Practice

Probability Distributions

For Exercises 1–3, the spinner shown is spun two times.

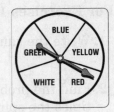

1. Write the sample space with all possible outcomes.

2. Find the probability distribution X, where X represents the number of times the spinner lands on blue for $X = 0$, $X = 1$, and $X = 2$.

3. Make a probability graph of the data.

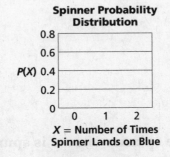

4. TELECOMMUNICATIONS Use the table that shows the probability distribution of the number of telephones per student's household at Wilson High.

X = Number of Telephones	1	2	3	4	5+
Probability	0.01	0.16	0.34	0.39	0.10

a. Show that the distribution is valid.

b. If a student is chosen at random, what is the probability that there are more than 3 telephones at the student's home?

c. Make a probability graph of the data.

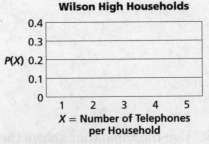

5. LANDSCAPING Use the table that shows the probability distribution of the number of shrubs (rounded to the nearest 50) ordered by corporate clients of a landscaping company over the past five years.

X = Number of Shrubs	50	100	150	200	250
Probability	0.11	0.24	0.45	0.16	0.04

a. Define a random variable and list its values.

b. Show that the distribution is valid.

c. What is the probability that a client's (rounded) order was at least 150 shrubs?

12-7 Skills Practice

Probability Simulations

1. CARDS Use a standard deck of 52 cards. Select a card at random, record the suit of the card (heart, diamond, club, or spade), and then replace the card. Repeat this procedure 26 times.

    a. Based on your results, what is the experimental probability of selecting a heart?

    b. Based on your results, what is the experimental probability of selecting a diamond or a spade?

    c. Compare your results to the theoretical probabilities.

2. SIBLINGS There are 3 siblings in the Bencievenga family. What could you use to simulate the genders of the 3 siblings?

3. TRANSPORTATION A random survey of 23 students revealed that 2 students walk to school, 12 ride the bus, 6 drive a car, and 3 ride with a parent or other adult. What could you use for a simulation to determine the probability that a student selected at random uses any one type of transportation?

4. BIOLOGY Stephen conducted a survey of the students in his classes to observe the distribution of eye color. The table shows the results of his survey.

Eye Color	Blue	Brown	Green	Hazel
Number	12	58	2	8

    a. Find the experimental probability distribution for each eye color.

    b. Based on the survey, what is the experimental probability that a student in Stephen's classes has blue or green eyes?

    c. Based on the survey, what is the experimental probability that a student in Stephen's classes does *not* have green or hazel eyes?

    d. If the distribution of eye color in Stephen's grade is similar to the distribution in his classes, about how many of the 360 students in his grade would be expected to have brown eyes?

12-7 Practice

Probability Simulations

1. MARBLES Place 5 red, 4 yellow, and 7 green marbles in a box. Randomly draw two marbles from the box, record each color, and then return the marbles to the box. Repeat this procedure 50 times.

  a. Based on your results, what is the experimental probability of selecting two yellow marbles?

  b. Based on your results, what is the experimental probability of selecting a green marble and a yellow marble?

  c. Compare your results to the theoretical probabilities.

2. OPTOMETRY Color blindness occurs in 4% of the male population. What could you use to simulate this situation?

3. SCHOOL CURRICULUM Laurel Woods High randomly selected students for a survey to determine the most important school issues among the student body. The school wants to develop a curriculum that addresses these issues. The survey results are shown in the table.

School Issues	
Issue	Number Ranking Issue Most Important
Grades	37
School Standards	17
Popularity	84
Dating	76
Violence	68
Drugs, including tobacco	29

  a. Find the experimental probability distribution of the importance of each issue.

  b. Based on the survey, what is the experimental probability that a student chosen at random thinks the most important issue is grades or school standards?

  c. The enrollment in the 9th and 10th grades at Laurel Woods High is 168. If their opinions are reflective of those of the school as a whole, how many of them would you expect to have chosen popularity as the most important issue?

  d. Suppose the school develops a curriculum incorporating the top three issues. What is the probability that a student selected at random will think the curriculum addresses the most important issue at school?